高 等 学 校 教 材

Organic Chemistry Experiments

有机化学实验

（双语版）

姜慧明　白日霞　吴小伟　刘世娟　主编

化学工业出版社

·北京·

内容简介

全书由6部分组成：有机化学实验应急要诀、有机化学实验基本技术、有机化合物合成通法、有机化学基本实验、有机化学系列实验、有机化学设计性实验。

本书可作为高等学校化学、应用化学、化学工程与工艺、生物工程、生物制药、食品化学、材料化学、环境工程、环境科学、生物技术等专业的有机化学实验课程教材，也可作为相关专业从业者的参考用书。

图书在版编目（CIP）数据

有机化学实验 ：双语版 ：汉英对照 / 姜慧明等主编. — 北京 ：化学工业出版社，2024.1

ISBN 978-7-122-45127-9

I. ①有… II. ①姜… III. ①有机化学－化学实验－高等学校－教材－汉、英 IV.①O62-33

中国国家版本馆 CIP 数据核字（2024）第 041989 号

责任编辑：李 琰 马 波 宋林青

责任校对：宋 玮 装帧设计：韩 飞

出版发行：化学工业出版社

（北京市东城区青年湖南街13号 邮政编码100011）

印 刷：三河市航远印刷有限公司

装 订：三河市宇新装订厂

787mm×1092mm 1/16 印张 11¾ 字数 291 千字

2024 年 5 月北京第 1 版第 1 次印刷

购书咨询：010-64518888 售后服务：010-64518899

网 址：http：//www.cip.com.cn

凡购买本书，如有缺损质量问题，本社销售中心负责调换。

定 价：35.00元

前 言

传统有机化学实验教学多以孤立地介绍单个有机化合物的制备方法为中心，着重强调的是实验操作技能的训练。本书以实践能力、创新能力培养为目标，从学习实验通法入手，着力将基本理论学习（基本实验）、综合能力训练（系列实验）、创新精神（设计性实验）和科研训练（双语实验）培养融为一体。本书仍以典型的有机反应为主线，将反应、合成、分离、提纯、物性测试等环节串联成一体。就选材的角度而言，主要从现代性、综合性、应用性及趣味性等方面来考虑。多数实验均以反应原理、实验通法、方案设计、合成实验、分离提纯等内容为主要教学环节，环环相扣，融成一体，借以加强实验教学的综合性训练，淡化其验证性色彩。其中，有些实验既可以作为单元反应实验的教学内容，又可作为多步合成实验教学中的一个环节，不仅节省了药品消耗，缩短了教学时间，而且增添了实验教学的研究性和探索性。

本书的创新与特色：一是引入了通法学习，在实验通法中还附有反应实例和参考文献，这不仅仅是为了展现实验通法的通用性和应用范围，更主要的还是为了拓宽学生的视野；二是引入了具有科研训练性质的双语实验内容，采取中英文对照的方式，切实提高了本科学生的专业英语的水平。

几年的教改实践表明，这种知识、能力、素质协调培养的教学改革不仅有利于学生学习和掌握有机反应的基本理论和实验技能，更有利于培养学生分析问题和解决问题的能力，尤其是有助于诱导学生发散思维，唤起他们的创新意识。

书中所选择的合成产物多数都具有实际应用价值，多年来我们一直致力于实验教学的改革和实践工作，并取得很好效果。

限于编者的水平，书中可能会有不妥之处，敬请使用本书的读者随时提出批评、指正。

编者

2023年11月于大连

目　录

第三部分　有机化合物合成通法　65

Section Ⅲ　General Method for Organic Compound Synthesis　86

第六部分　有机化学设计性实验 167

Section Ⅵ　Organic Chemistry Design-Oriented Experiments 172

参考文献 178

第一部分

有机化学实验应急要诀

- 如果出现危险，应立刻报告指导教师。
- 如果着火，切勿惊慌。当衣服着火时，可用浸湿的工作服将着火部位裹起来，或者直接用水冲淋；如果烧杯或烧瓶中的试剂着火，首先应关灭火源，然后用石棉网或湿抹布覆盖瓶口将火熄灭；若遇大火，就要使用灭火器。
- 如果烫伤，可涂抹烫伤油膏，如蓝油烃油膏。
- 如果酸灼伤，立刻用水冲洗，然后用1%碳酸氢钠水溶液洗涤，经水冲洗后涂上凡士林。
- 如果碱灼伤，立刻用水冲洗，然后用1%硼酸水溶液洗涤，经水冲洗后涂上凡士林。
- 如果溴灼伤，立刻用水冲洗，然后用酒精擦洗，再涂上甘油轻轻按摩。
- 如果试剂溅入眼睛，立刻用水冲洗。
- 如果割伤，将伤口处异物（如玻璃屑）取出，用水冲洗伤口，涂上红汞药水后用纱布包扎。伤势严重者应马上送医院就医。
- 在使用贴有危险品警示图标的药品和试剂时，要特别小心谨慎。常见危险品警示图标如下：

Emergency Tips for Organic Chemistry Experiments

- If there is a danger, immediately report to the supervising teacher.
- In case of fire, remain calm. If clothing catches fire, use a damp lab coat to wrap around the affected area or directly douse it with water. If the reagents in a beaker or flask catch fire, first extinguish the ignition source, then cover the mouth of the container with asbestos mesh or a damp cloth to put out the fire. In the event of a large fire, use a fire extinguisher.
- For burns, apply burn ointment, such as petroleum jelly or blue ointment.
- For acid burns, rinse immediately with water, then wash with a 1% sodium bicarbonate solution, followed by another water rinse. Apply petroleum jelly after washing.
- For alkali burns, rinse immediately with water, then wash with a 1% boric acid solution, followed by another water rinse. Apply petroleum jelly after washing.
- For bromine burns, rinse immediately with water, then rub with alcohol and gently massage with glycerin.
- If reagents splash into the eyes, immediately rinse with water.
- For cuts, remove any foreign objects (such as glass fragments) from the wound, rinse the wound with water, apply mercuric oxide solution, and bandage with gauze. For severe injuries, seek immediate medical attention.
- Exercise extra caution when handling drugs and reagents labeled with hazardous material warning symbols. Common hazardous material warning symbols include:

CORROSIVE

IRRITANT

HAZARD

第二部分

有机化学实验基本技术

2.1 常用玻璃仪器及实验装置

在有机实验中经常会用到一些玻璃仪器及实验装置，熟悉这些仪器、装置及其维护方法是十分必要的。

2.1.1 常用玻璃仪器

常用的玻璃仪器分为两类：普通玻璃仪器和标准磨口仪器。

标准磨口，顾名思义，接口部位的尺寸大小都是统一的，即标准化的。例如，14 口、19 口、24 口指的就是磨口的最大端直径分别为 14 mm、19 mm 和 24 mm。只要是相同尺寸的标准磨口，相互之间便可以装配吻合。对不同尺寸的磨口仪器，还可以通过相应尺寸的大小磨口接头使之相互连接。

在使用标准磨口仪器的过程中应该注意，装配时要对齐，不可用力过猛，以免破裂。一般情况下，磨口处不必涂润滑剂。若作减压蒸馏时，应适当地涂抹真空脂。实验结束后应及时拆卸仪器，以免黏结难卸。常用标准磨口仪器及其他玻璃仪器见图 2-1～图 2-3。

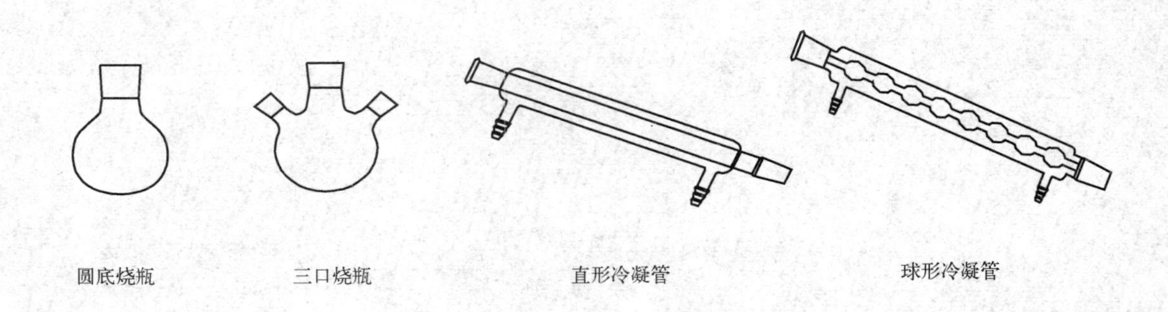

圆底烧瓶　　三口烧瓶　　直形冷凝管　　球形冷凝管

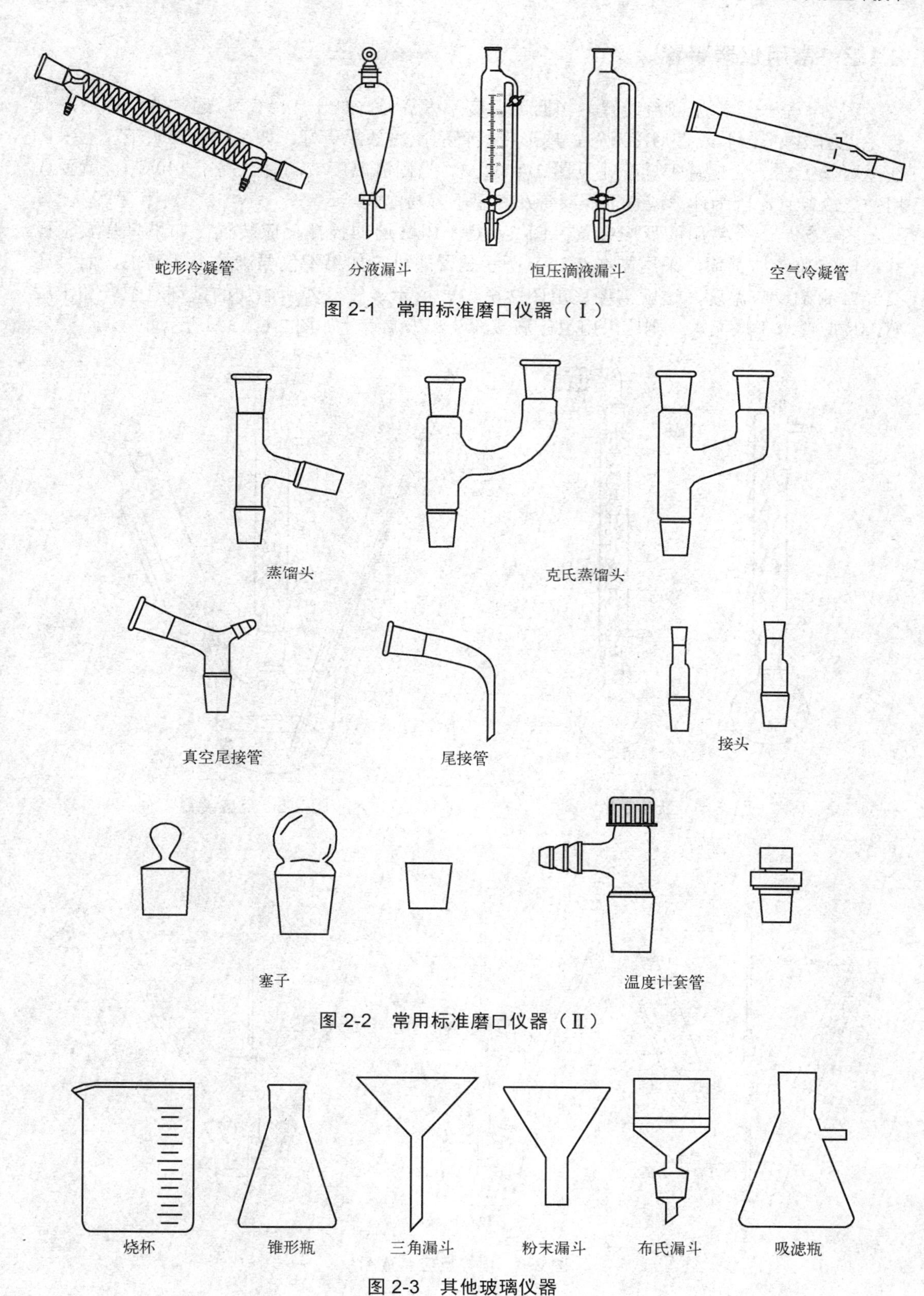

图 2-1　常用标准磨口仪器（Ⅰ）

图 2-2　常用标准磨口仪器（Ⅱ）

图 2-3　其他玻璃仪器

2.1.2 常用仪器装置

图 2-4 是一组常见回流装置。当回流温度不太高时（低于 140℃），通常选用球形冷凝管［见图 2-4（1)］或直形冷凝管，球形冷凝管的冷凝效果更好。如果实验要求干燥无水，在冷凝管上端还应配置干燥管［见图 2-4（2)］。当回流温度较高时（高于 140℃），就要选用空气冷凝管，因为球形或直形冷凝管在高温下容易炸裂。

图 2-5 是一个常用的反应装置。图 2-6 是一组常用的搅拌反应装置。如果只是要求搅拌、回流和滴加试剂，采用图 2-6（1）所示装置即可。如果不仅要满足上述要求，而且还要经常测试反应温度，这就需要采用图 2-6（2）所示装置，若所添加的试剂对空气和水敏感，反应要求干燥无水，则应采用恒压滴液漏斗和干燥管［如图 2-6（3）所示］。

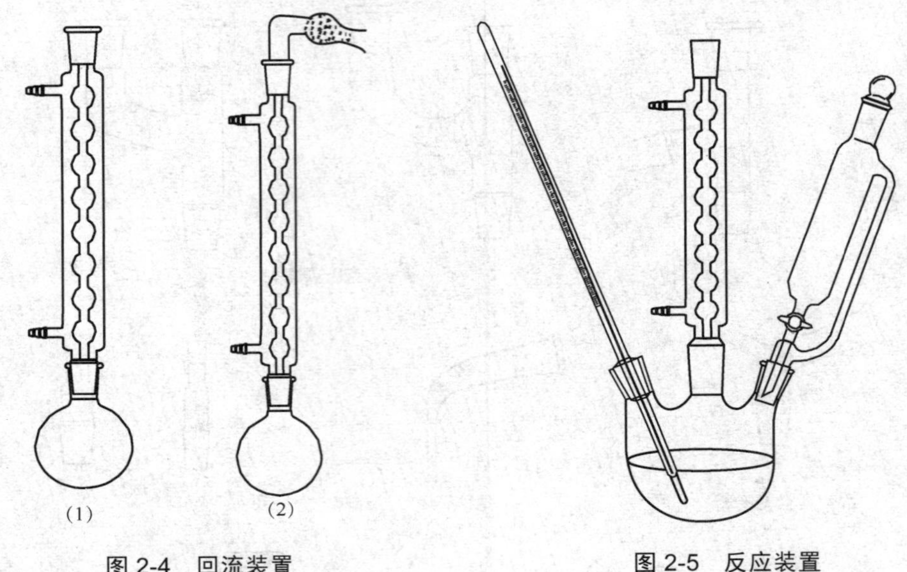

图 2-4 回流装置

图 2-5 反应装置

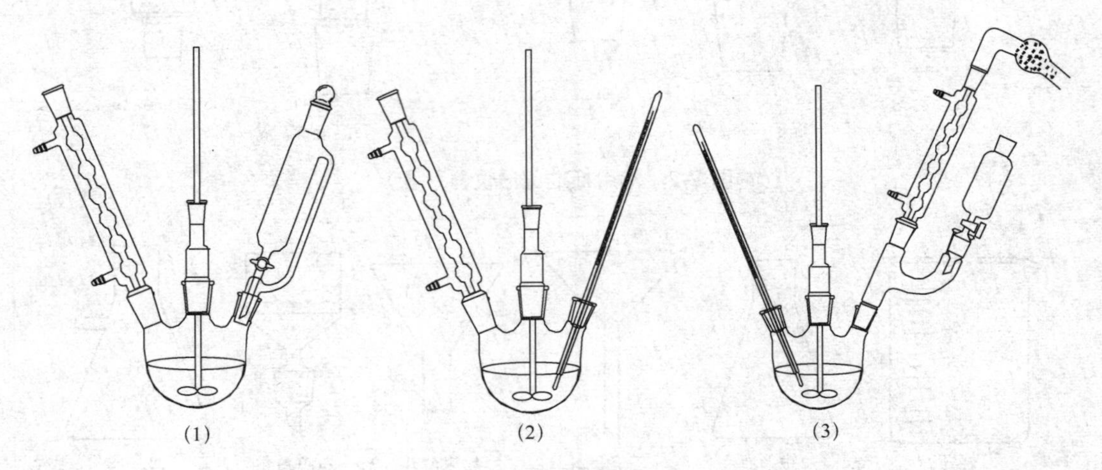

图 2-6 搅拌反应装置图

图 2-7 是一组气体吸收装置。在烧杯中装入气体吸收液，以吸收反应过程中产生的碱性或酸性气体。

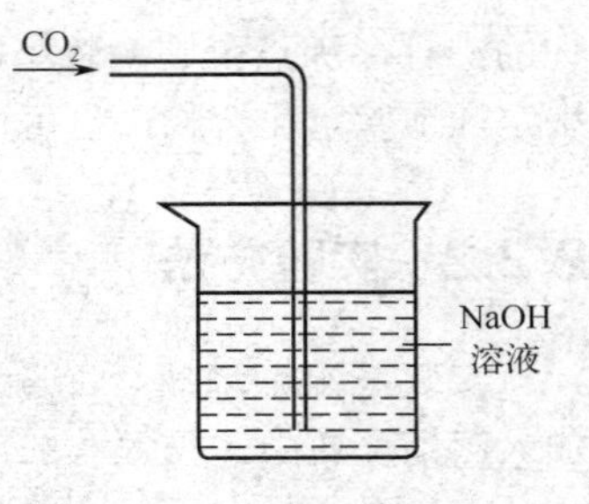

图 2-7　气体吸收装置

2.1.3　注意事项

（1）两个磨口仪器如果黏结在一起，不可盲目使劲拆卸，可先用电吹风对着黏结接口处加热，然后再试着拆卸。

（2）对玻璃仪器加热时，除了试管外一般都不可直接用火加热，以防破裂。

（3）厚壁玻璃仪器（如吸滤瓶）受热易破裂，故不可直接对其加热。计量类容器（如量筒）受热会影响计量准确度，洗净后宜晾干而不宜置于高温下烘烤。

（4）具塞玻璃仪器（如滴液漏斗）不用时，应该将旋塞与磨口之间用纸片隔离开来，以免粘牢。

（5）用玻璃仪器盛装碱性溶液时，使用完后应及时洗涤，以防止黏结。

（6）清洗玻璃仪器时，可用去污粉或者家用洗洁精进行洗涤，清洗完毕，用清水冲净，倒置在玻璃仪器架上晾干。如果需要快速晾干，可用少许丙酮或乙醇进行涮洗，洗毕用电吹风干燥。

（7）在回流装置中，多采用球形冷凝管。因为蒸气与冷凝管接触面积较大，冷凝效果较好，尤其适用于低沸点溶剂的回流操作。当回流温度高于 150℃时，选用空气冷凝管。

（8）在搅拌反应中，如果反应混合物量较大，或较黏稠，或含有固体物质，这时，用磁力搅拌效果不佳，应以机械搅拌器搅拌为宜。

（9）在采用气体吸收装置（见图 2-7）时，应密切注意观察气体吸收情况。有时会因为反应温度的变化而在体系内形成一定的负压，从而发生气体吸收液倒吸现象，需保持玻璃漏斗或玻璃管悬在液面上，使反应体系与大气相通，消除负压。

2.2　蒸馏

液态物质受热沸腾成蒸气，蒸气经冷凝又转变为液体，这个操作过程就称作蒸馏（Distillation）。蒸馏是纯化和分离液态物质的一种常用方法，通过蒸馏还可以测定纯液态物质的沸点。

2.2.1　实验原理

纯的液态物质在一定压力下具有确定的沸点，不同的物质具有不同的沸点。蒸馏操作利用不同物质的沸点差异对液态混合物进行分离和纯化。当液态混合物受热时，由于低沸点物质易挥发，首先被蒸出，而高沸点物质因不易挥发或挥发出的少量气体易被冷凝而滞留在蒸馏瓶中，从而使混合物得以分离。不过，只有当组分沸点相差在 30℃以上时，蒸馏才有较好的分离效果。如果组分沸点差异不大，就需要采用分馏操作对液态混合物进行分离和纯化。

需要指出的是，具有恒定沸点的液体并非都是纯化合物，因为有些化合物相互之间可以形成二元或三元共沸混合物，而共沸混合物是不能通过蒸馏操作进行分离的。通常，纯化合物的沸程（沸点范围）较小（约 0.5～1℃），而混合物的沸程较大。因此，蒸馏操作既可用来定性地鉴定化合物，也可用以判定化合物的纯度。

2.2.2 实验方法

安装好蒸馏烧瓶、冷凝管、接引管和接收瓶（见图 2-8），然后将待蒸馏液体通过漏斗加入蒸馏烧瓶中，投入 1～2 粒沸石，再配置温度计。

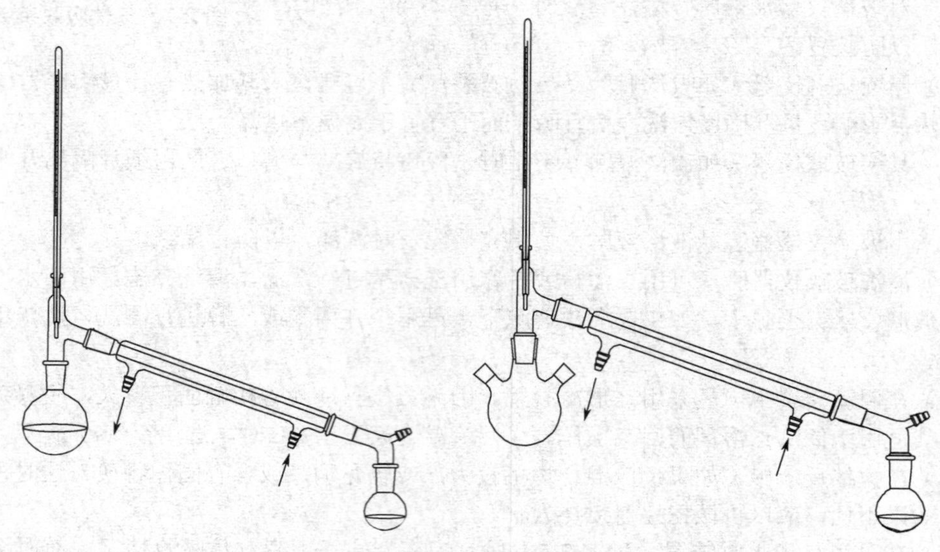

图 2-8 简单蒸馏装置图

接通冷凝水，开始加热，使瓶中液体沸腾。调节火焰，控制蒸馏速度，以 1～2 滴/秒为宜。在蒸馏过程中，注意温度计读数的变化，记下第一滴馏出液流出时的温度。当温度计读数稳定后，另换一个接收瓶收集馏分。如果仍然保持平稳加热，但不再有馏分流出，而且温度会突然下降，这表明该段馏分已近蒸完，需停止加热，记下该段馏分的沸程和体积（或质量）。馏分的温度范围愈小，其纯度就愈高。

有时，在有机反应结束后，需要对反应混合物进行直接蒸馏，此时，可以将三口烧瓶作蒸馏瓶组装成蒸馏装置直接进行蒸馏（见图 2-9）。

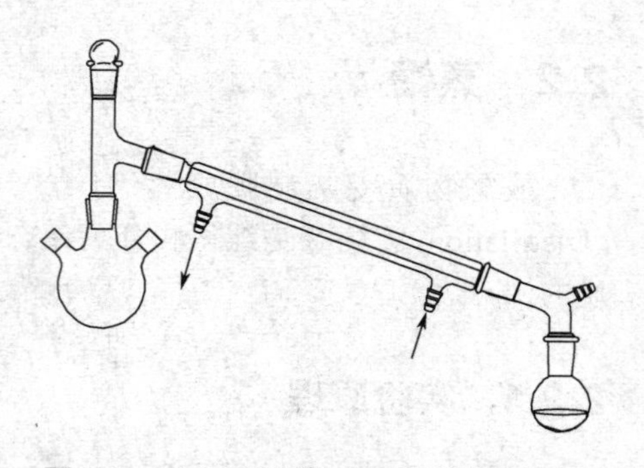

图 2-9 由反应装置改装的蒸馏装置图

2.2.3 注意事项

（1）蒸馏烧瓶的大小依蒸馏液体的量而定。通常，待蒸馏液体的体积约占蒸馏烧瓶体积的 1/3～2/3。

（2）当待蒸馏液体的沸点在 140℃以下时，应选用直形冷凝管；沸点在 140℃以上时，就要选用空气冷凝管，若仍用直形冷凝管则易发生爆裂。

（3）如果蒸馏装置中所用的接引管无侧管，则接引管和接收瓶之间应留有空隙，以确保蒸馏装置与大气相通。否则，封闭体系受热后会引发事故。

（4）沸石是一种多孔性的物质，如素瓷片或毛细管。当液体受热沸腾时，沸石内的小气泡就成为气化中心，使液体保持平稳沸腾。如果蒸馏已经开始，但忘了投沸石，此时千万不要直接投放沸石，以免引发暴沸。正确的做法是，先停止加热，待液体稍冷片刻后再补加沸石。

（5）蒸馏低沸点易燃液体（如乙醚）时，千万不可用明火加热，此时可用热水浴加热。在蒸馏沸点较高的液体时，可以用明火加热。明火加热时，烧瓶底部一定要放置石棉网，以防因受热不匀而炸裂。

（6）无论何时，都不要使蒸馏烧瓶蒸干，以防意外。

2.3　分馏

简单蒸馏只能对沸点差异较大的混合物作有效的分离，而采用分馏柱进行蒸馏则可对沸点相近的混合物进行分离和提纯，这种操作方法称为分馏（Fractional Distillation）。简单地说，分馏就是多次蒸馏，利用分馏技术甚至可以将沸点相距1～2℃的混合物分离开来。

2.3.1　实验原理

当混合物受热沸腾时，其蒸气首先进入分馏柱。由于柱内外存在温差，柱内蒸气中高沸点组分受柱外空气的冷却而被冷凝，并流回至烧瓶，从而导致继续上升的蒸气中低沸点组分的含量相对增加。这一个过程可以看作是一次简单的蒸馏。当高沸点冷凝液在回流途中遇到新蒸上来的蒸气时，两者之间发生热交换，上升的蒸气中，同样是高沸点组分被冷凝，低沸点组分继续上升。这又可以看作是一次简单蒸馏。蒸气就是这样在分馏柱内反复地进行着气化、冷凝和回流的过程，或者说，重复地进行着多次简单蒸馏。因此，只要分馏柱的效率足够高，从分馏柱上端蒸出的蒸气组分就能接近低沸点单组分的纯度，而高沸点组分仍回流到蒸馏烧瓶中。需要指出的是，由于共沸混合物具有恒定的沸点，与蒸馏一样，分馏操作也不可用来分离共沸混合物。

2.3.2　实验方法

将待分馏物质装入圆底烧瓶，并投放几粒沸石，然后依序安装分馏柱、温度计、冷凝管、接引管及接收瓶（见图2-10）。

接通冷凝水，开始加热，使液体平稳沸腾。当蒸气缓缓上升时，注意控制温度，使馏出速度维持在2～3秒钟一滴。记录第一滴馏出液滴入接收瓶时的温度，然后根据具体要求分段收集馏分，并记录各馏分的沸点范围及体积。

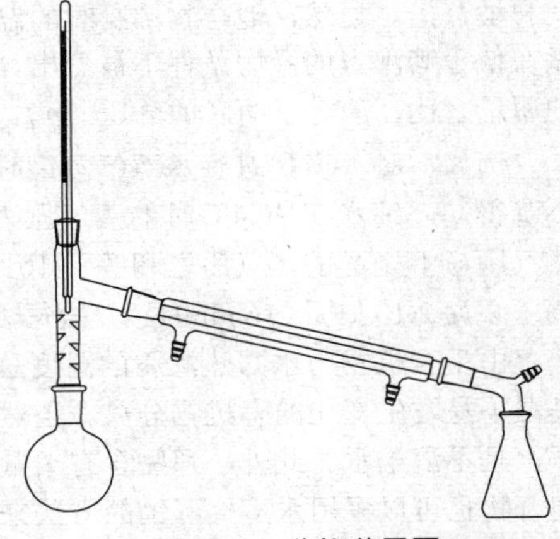

图2-10　分馏装置图

2.3.3 注意事项

（1）分馏柱柱高是影响分馏效率的重要因素之一。一般来讲，分馏柱越高，上升蒸气与冷凝液间的热交换次数就越多，分离效果就越好。但是，如果分馏柱过高，也会影响馏出速度。

（2）分馏柱内的填充物也是影响分馏效率的一个重要因素。填充物在柱中起到增加蒸气与回流液接触的作用，填充物比表面积越大，越有利于提高分离效率。不过，需要指出的是，填充物之间要保持一定的空隙，否则会导致蒸馏困难。实验室中常用的韦氏（Vigreux）分馏柱是一种柱内呈刺状的简易分馏柱，不需另加填料。

（3）当室温较低或待分馏液体的沸点较高时，分馏柱的绝热性能就会对分馏效率产生显著影响。在这种情况下，如果分馏柱的绝热性能差，其散热就快，因而难以维持柱内气液两相间的热平衡，从而影响分离效果。为了提高分馏柱的绝热性能，可用玻璃布等保温材料将柱身裹起来。

（4）在分馏过程中，要注意调节加热温度，使馏出速度适中。如果馏出速度太快，就会产生液泛现象，即回流液来不及流回至烧瓶，并逐渐在分馏柱中形成液柱。若出现这种现象，应停止加热，待液柱消失后重新加热，使气液达到平衡，再恢复收集馏分。

2.4 水蒸气蒸馏

将水蒸气通入不溶于水的有机物中或使有机物与水经过共沸而蒸出有机物，这个操作过程称为水蒸气蒸馏（Steam Distillation）。水蒸气蒸馏是分离和提纯液态或固态有机物的一种方法。

2.4.1 实验原理

根据道尔顿分压定律，当水与有机物混合共热时，其蒸气压为各组分之和。即

$$p_{\text{混合物}}=p_{\text{水}}+p_{\text{有机物}}$$

如果水的蒸气压和有机物的蒸气压之和等于大气压，混合物就会沸腾，有机物和水就会一起被蒸出。显然，混合物沸腾时的温度要低于其中任一组分的沸点。换句话说，有机物可以在低于其沸点的温度条件下被蒸出。从理论上讲，馏出液中有机物（$m_{\text{有机物}}$）与水（$m_{\text{水}}$）的质量之比，应等于两者的分压（$p_{\text{有机物}}$和$p_{\text{水}}$）与各自分子量（$M_{\text{有机物}}$和$M_{\text{水}}$）乘积之比。

例如，对1-辛醇进行水蒸气蒸馏时，1-辛醇与水的混合物在99.4℃沸腾。通过查阅手册不难得知，纯水在99.4℃时的蒸气压力99.18 kPa（744 mmHg）。按道尔顿分压定律，水的蒸气压与1-辛醇的蒸气压之和等于101.31 kPa（760 mmHg）。因此，1-辛醇在99.4℃时的蒸气压必为2.13 kPa（16 mmHg）。故每蒸出1 g水便有0.16 g 1-辛醇被蒸出。

由于有机物与水共热沸腾的温度总在100℃以下，因此，水蒸气蒸馏操作特别适用于在高温下易发生变化的有机物分离。当然，有机物还须具有至少为0.7 kPa（5 mmHg）的蒸气压，且不溶于水。此外，那些含有大量树脂状杂质、直接用蒸馏或重结晶等方法难以分离的混合物也可以采用水蒸气蒸馏的方法来分离。

2.4.2　实验方法

依序安装水蒸气发生器、圆底烧瓶、克氏蒸馏头、温度计、冷凝管、接引管和接收瓶［参见图 2-11（1）］。将待分离混合物转入烧瓶中，将 T 形管活塞打开，加热水蒸气发生器使水沸腾。当有水蒸气从 T 形管支口喷出时，关闭支管口，使水蒸气通入烧瓶。连通冷却水，使混合蒸气能在冷凝管中迅速冷凝而流入接收瓶。馏出速度以 2 滴/秒为宜，通过调节火焰加以控制。当馏出液清亮透明、不再含有油状物时，即可停止蒸馏。先打开 T 形管支口，然后停止加热。将收集液转入分液漏斗，静置分层，除去水层，即得分离产物。

如果不用水蒸气发生器而采用一种更为简单的水蒸气蒸馏装置也可以正常地进行水蒸气蒸馏操作［见图 2-11（2）］。先将待分离有机物和适量的水置入圆底烧瓶中，再投入几粒沸石，接通冷凝水，开始加热，保持平稳沸腾。其他操作同前，只是当烧瓶内的水经连续不断地蒸馏而减少时，可通过蒸馏头上配置的滴液漏斗补加水。如果依装置图 2-11（2）进行水蒸气蒸馏操作容易使混合物溅入冷凝管，使分离纯化受到影响，那么采用图 2-11（3）来操作就可以有效地避免这个问题。不过，由于克氏蒸馏头弯管段较长，蒸气易冷凝，影响有效蒸馏。此时，可以用玻璃棉等绝热材料缠绕，以避免热量迅速散失，从而提高蒸馏效率。

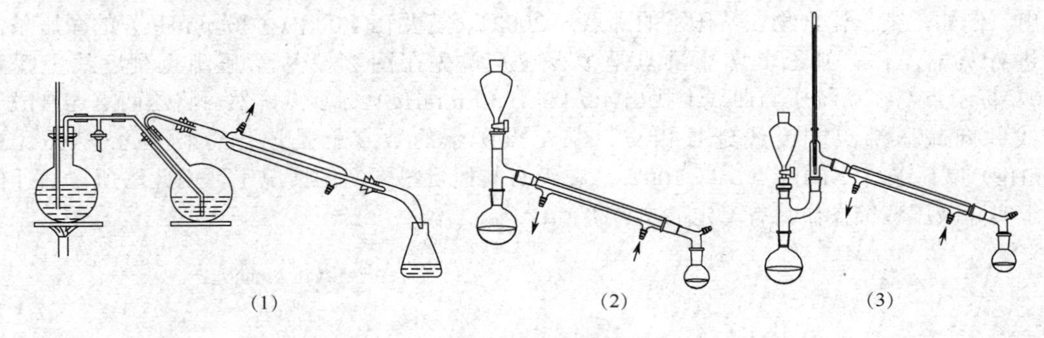

图 2-11　水蒸气蒸馏

2.4.3　注意事项

（1）水蒸气发生器中一定要配置安全管。可选用一根长玻璃管作安全管，管子下端要接近水蒸气发生器底部。使用时，注入的水不要过多，一般不要超出其容积的 2/3。

（2）水蒸气发生器与烧瓶之间的连接管路应尽可能短，以减少水蒸气在导入过程中的热损耗。

（3）导入水蒸气的玻璃管应尽量接近圆底烧瓶底部，以利提高蒸馏效率。

（4）在蒸馏过程中，如果有较多的水蒸气因冷凝而积聚在圆底烧瓶中，可以用小火隔着石棉网在圆底烧瓶底部加热。

（5）实验中，应经常注意观察安全管。如果其中的水柱出现不正常上升，应立即打开 T 形管，停止加热，找出原因，排除故障后再重新蒸馏。

（6）停止蒸馏时，一定要先打开 T 形管，然后停止加热。如果先停止加热，水蒸气发生器因冷却而产生负压，会使烧瓶内的混合液发生倒吸。

2.5 减压蒸馏

有些有机化合物热稳定性较差，常常在受热温度还未到达其沸点就已发生分解、氧化或聚合。对这类化合物的纯化或分离就不宜采取常压蒸馏的方法而应该在减压条件下进行蒸馏。减压蒸馏又称真空蒸馏（Vacuum Distillation），可以将有机化合物在低于其沸点的温度下蒸馏出来。减压蒸馏尤其适用于蒸馏那些沸点高、热稳定性差的有机化合物。

2.5.1 实验原理

液体化合物的沸点与外界压力有密切的关系。当外界压力降低时，使液体表面分子逸出而沸腾所需要的能量也会降低。换句话说，如果降低外界压力，液体沸点就会随之下降。例如，苯甲醛在常压下的沸点为 179℃/101.3 kPa（760 mmHg），当压力降至 6.7 kPa（50 mmHg）时，其沸点已降低到 95℃。通常，当压力降低到 2.67 kPa（20 mmHg）时，多数有机化合物的沸点要比其常压下的沸点低 100℃左右。沸点与压力的关系可近似地用图 2-12 推出。例如，某一化合物在常压下的沸点为 200℃，若要在 4.0 kPa（30 mmHg）的减压条件下进行蒸馏操作，那么其蒸出沸点是多少呢？首先在图 2-12 中常压沸点刻度线上找到 200℃标示点，在系统压力曲线上找出 4.0 kPa（30 mmHg）标示点，然后将这两点连接成一直线并向减压沸点刻度线延长相交，其交点所示的数字就是该化合物在 4.0 kPa（30 mmHg）减压条件下的沸点，即 100℃。在没有其他资料来源的情况下，由此法所得估计值对于实际减压蒸馏操作具有一定的参考价值。

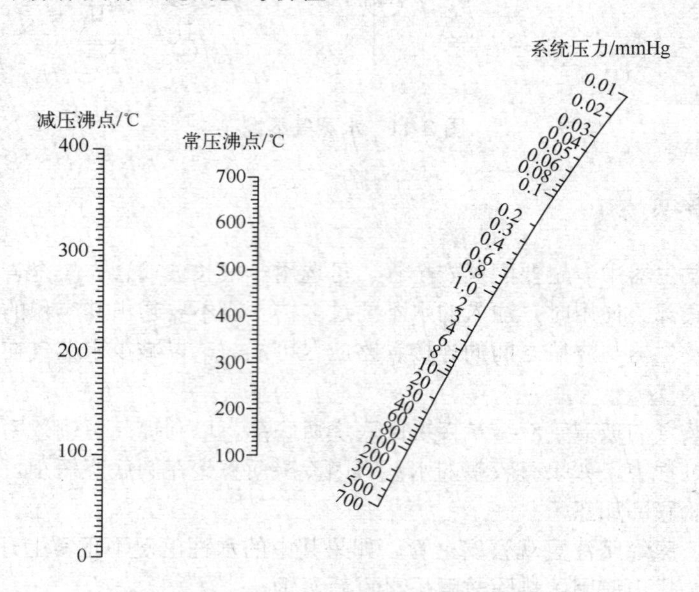

图 2-12 液体在常压和减压下的沸点近似关系图

2.5.2 实验方法

通常，减压蒸馏系统是由蒸馏装置、安全瓶、气体吸收装置、缓冲瓶及测压装置组成。

在作减压蒸馏操作时，依次装配蒸馏烧瓶、克氏蒸馏头、冷凝管、真空接引管及接收瓶，以玻璃漏斗将待蒸馏物质注入蒸馏烧瓶中，配置毛细管，使毛细管尽量接近瓶底［见图 2-13（1)］。

将真空接引管用厚壁真空橡皮管依序与安全瓶、冷却阱、真空计、气体吸收塔、缓冲瓶及油泵相连接（见图 2-14）。冷却阱可置于广口保温瓶中，用液氮或冰-盐冷却剂冷却。

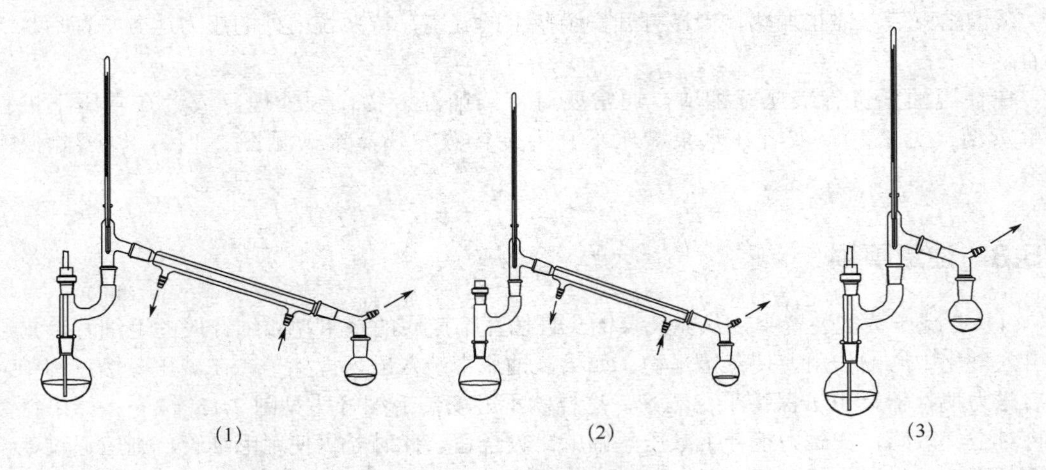

图 2-13　减压蒸馏装置

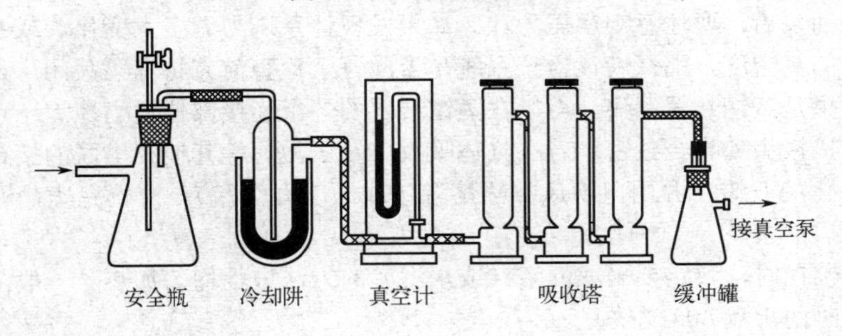

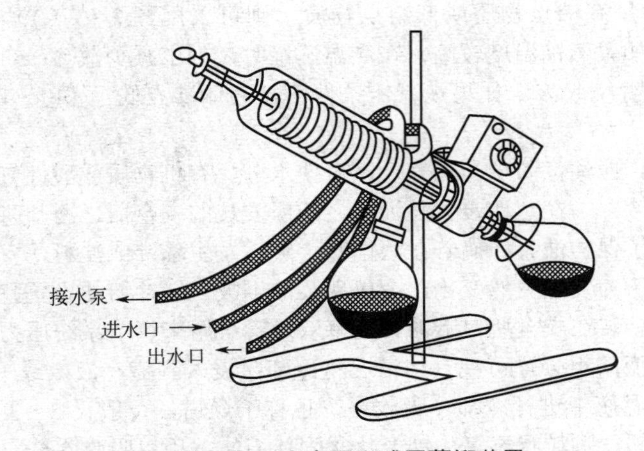

图 2-14　真空泵和减压蒸馏装置

先打开安全瓶上的活塞，使体系与大气相通。然后开启油泵抽气，慢慢关闭安全瓶上的旋塞，同时注意观察压力计读数的变化。通过小心旋转安全瓶上的旋塞，使体系真空度调节至所需值。

接通冷凝管上的冷凝水，开始用热浴液对蒸馏烧瓶加热，通常浴液温度要高出待蒸馏物质减压时的沸点30℃左右。蒸馏速度以1～2滴/秒为宜。当有馏分蒸出时，记录其沸点及相应的压力读数。如果待蒸馏物中有几种不同沸点的馏分，可通过旋转多头接引管、收集不同的馏分。

蒸馏结束后，停止加热，慢慢打开安全瓶上的旋塞，待系统内外的压力达到平衡后，关闭油泵。

在使用油泵进行减压蒸馏前，通常要对待蒸馏混合物作预处理，或者在常压下进行简单蒸馏（见2.2），或者在水泵减压下利用旋转蒸发仪蒸馏（见图2-14），以蒸除低沸点组分。

2.5.3 注意事项

（1）在减压蒸馏装置中，从克氏蒸馏头直插蒸馏瓶底的是末端如细针般的毛细管，它起到引入气化中心的作用，使蒸馏平稳。如果蒸馏瓶中装入磁力搅拌子，在减压蒸馏过程中，开启磁力搅拌器，也可保持平稳蒸馏，这样就不必安装毛细管［见图2-13（2）］。如果待蒸馏物对空气敏感，在磁力搅拌下减压蒸馏就比较合适。此时若仍使用毛细管，则应通过毛细管导入惰性气体（如氮气），来加以防护。

（2）打开油泵后，要注意观察压力计。如果发现体系压力无多大变化，或系统不能达到油泵应该达到的真空度，那么就该检查系统是否漏气。检查前先将油泵关闭，再分段查那些连接部位。如果是蒸馏装置漏气，可以在蒸馏装置的各个连接部位适当地涂一点真空脂，并通过旋转使磨口接头处吻合致密。若在气体吸收塔及压力计等其他相串联的接合部位漏气，可涂上少许熔化的石蜡，并用电吹风加热熔融（或涂上真空脂）。检查完毕，即可按实验方法所述程序开启油泵。

（3）减压蒸馏时，一定要采取油浴（或水浴）的方法进行均匀加热。一般浴温要高出待蒸馏物在减压时的沸点30℃左右。

（4）如果蒸馏少量高沸点物质或低熔点物质，则可采用图2-13（3）所示装置进行蒸馏，即省去冷凝管。如果蒸馏温度较高，在高温蒸馏时，为了减少散热，可在克氏蒸馏头处用玻璃棉等绝热材料缠绕起来。如果在减压条件下，液体沸点低于140～150℃，可用冷水浴对接收瓶冷却。

（5）使用油泵时，应注意防护与保养，不可使水分、有机物质或酸性气体侵入泵内，否则会严重降低油泵的效率。在蒸馏装置与油泵之间所安装的安全瓶、冷却阱、气体吸收塔及缓冲瓶，目的就是为了保护油泵。倘若在蒸馏时，突然发生暴沸或冲料，安全瓶就起到防护作用。有时，系统内压力发生突然变化，会出现泵油倒吸，缓冲瓶的设置就可以避免泵油冲入气体吸收塔。另外，装在安全瓶口上的带旋塞双通管可用来调节系统压力或放气。对于那些被抽出来的沸点较低的组分，可视具体情况将冷却阱浸入到盛有液氮或干冰或冰-水或冰-盐等冷却剂的广口保温瓶中进行冷却。吸收塔，也称干燥塔，一般设2～3个。这些干燥塔中分别装有无水氯化钙、颗粒状氢氧化钠及片状固体石蜡，用以吸收水分、酸性气体及烃类气体。应该指出的是，在用油泵减压蒸馏前，一定要先作简单蒸馏或用水泵减压蒸馏，以蒸除低沸点物质，防止低沸点物质抽入油泵。

（6）图2-15为封闭式水银压力计，常用于测量减压系统的真空度，其两臂汞面高度之差即为减压系统的真空度。使用时应当注意，当减压操作结束时，要小心旋开安全瓶上的双通旋塞，让气体慢慢进入系统，使压力计中的水银柱缓缓复原，以避免因系统内的压力突增使水银柱冲破玻璃管。

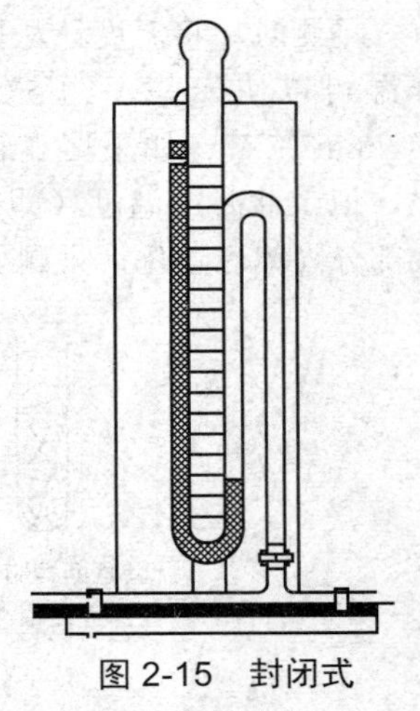

图 2-15　封闭式水银压力计

2.6　熔点测定

在大气压力下，化合物受热由固态转化为液态时的温度称为该化合物的熔点。熔点是固体有机化合物的物理常数之一，通过测定熔点不仅可以鉴别不同的有机化合物，而且可以判断其纯度。

2.6.1　实验原理

严格地说，所谓熔点指的是在大气压力下化合物的固-液两相达到平衡时的温度。通常纯的有机化合物都具有确定的熔点，而且从固体初熔到全熔的温度范围（称熔程或熔距）很窄，一般不超过 0.5～1℃。但是，如果样品中含有杂质，就会导致熔点下降、熔距变宽。因此，通过测定熔点，观察熔距，可以很方便地鉴别未知物，并判断其纯度。显然，这一性质可用来鉴别两种具有相近或相同熔点的化合物究竟是否为同一化合物。方法十分简单，只要将这两种化合物混合在一起，并观测其熔点。如果熔点下降，而且熔距变宽，那必定是两种性质不同的化合物。需要指出的是，有少数化合物，受热时易发生分解。因此，即使其纯度很高，也不具有确定的熔点，而且熔距较宽。

2.6.2　实验方法

将干燥过研细的待测样品放置在干燥洁净的表面皿上，然后用测熔点毛细管开口的一端垂直插入粉末状的样品中，即见有些许样品进入毛细管。再将毛细管开口端朝上，让毛细管封口端在实验台上轻击几下，样品便落入毛细管底部。如此反复操作几次，然后让毛细管封口端朝下，在一长约 50 cm 直立于表面皿上的玻璃管中自由落下，反复操作几次，使毛细管中的样品装得致密均匀；样品高约 4 mm。然后将装有样品的毛细管用细橡皮圈固定在温度计上，并使毛细管装样部位位于水银球处（见图 2-16）。

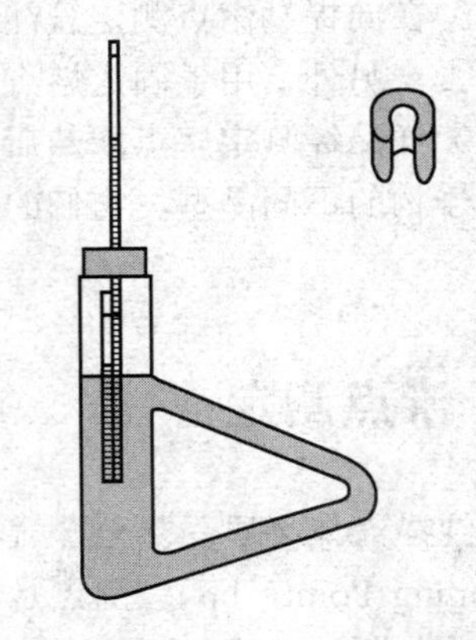

图 2-16　提勒（Thiele）熔点测定管

将提勒（Thiele）熔点测定管固定在铁架台上，注入导热液，使导热液液面位于提勒熔点测定管交叉口处。管口配置开有小槽的软木塞，将带有测熔点毛细管的温度计插入其中，使温度计的水银球位于提勒熔点测定管两支管的中间。

粗测时，用小火在提勒熔点测定管底部加热，升温速度以 5℃/min 为宜。仔细观察温度的变化及样品是否熔化。记录样品熔化时的温度，即得试样的粗测熔点。移去火焰，让导热液温度降至粗测熔点以下约 30℃，即可参考粗测熔点进行精测。

精测时，将温度计从提勒熔点测定管中取出，换上第二根熔点管后便可加热测定。初始升温可以快一些，约5℃/min；当温度升至离粗测熔点约10℃时，要控制升温速度在1℃/min左右。如果熔点管中的样品出现塌落、湿润，甚至显现出小液滴，即表明开始熔化，记录此时的温度（即初熔温度）。继续缓缓升温，直至样品全熔，记录全熔（即管中绝大部分固体已熔化，只剩少许即将消失的细小晶体）时的温度。固体熔化过程参见图2-17。

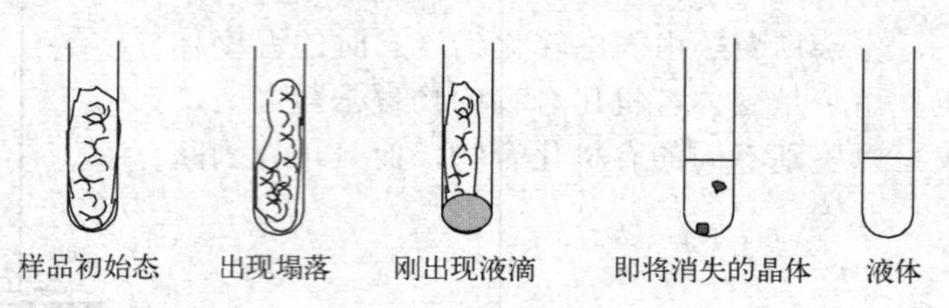

图2-17 固体熔化过程

2.6.3 注意事项

（1）用提勒熔点测定管测定熔点是实验室中常用的一种测定熔点的方法。此外，还可采用显微熔点测定仪或数字熔点仪。其中，用显微熔点测定仪测定熔点具有使用样品少、可测高熔点样品、可观察样品在受热过程中的变化等特点。

（2）待测样品一定要经充分干燥后再进行测定熔点。否则，含有水分的样品会导致其熔点下降、熔距变宽。另外，样品还应充分研细，装样要致密均匀，否则，样品颗粒间传热不匀，也会使熔距变宽。

（3）导热介质的选择可根据待测物质的熔点而定。若熔点在95℃以下，可以用水作导热液；若熔点在95～220℃范围内，可选用液体石蜡油；若熔点温度再高些，可用浓硫酸（250～270℃），但需注意安全。

（4）在向提勒熔点测定管注入导热液时不要过量。要考虑到导热液受热后，其体积会膨胀的因素。另外，用于固定熔点管的细橡皮圈不要浸入导热液中，以免溶胀脱落。

（5）样品经测定熔点冷却后又会转变为固态，由于结晶条件不同，会产生不同的晶型。同一化合物的不同晶型，它们的熔点常常不一样。因此，每次测熔点都应该使用新装样品的熔点管。

2.7 沸点测定

当纯净液体物质受热至蒸气压与外界压力相等时就会沸腾，此时的温度就是该物质的沸点（Boiling Point，bp）。沸点是有机化合物的物理常数之一，通过测定沸点可以鉴别有机化合物，并判断其纯度。

2.7.1 实验原理

在一定压力下，每一种化合物都有其特定的沸点。换句话说，同一种化合物在不同的压力下，其沸点是不同的。因此，描述一种化合物的沸点常要注明其压力条件。例如，二苯甲酮在13.3 kPa（100 mmHg）时，沸腾温度为224.4℃，记为224.4℃/13.3 kPa。不过，如果指的是其常压下的沸点，则通常不注明压力条件。例如，二苯甲酮的沸点记为305.4℃，指的就是常压下的沸点。需要指出的是，具有恒定沸点的液体并不一定都是纯化合物，因为共沸混合物也具有恒定的沸点。因此，测定沸点只能定性地鉴别一个化合物。不过，就一种已知物而言，通过测其沸点，看其沸程范围是可以判断其纯度的。因为，纯化合物的沸程一般较窄，约为0.5～1℃。测定沸点的方法有常量法和微量法。常量法采用的是蒸馏装置，其方法与简单蒸馏操作相同（见2.2）；而微量法所使用的装置与熔点测定装置相似。

2.7.2　实验方法

以内径3～4 mm、长8～10 cm、一端封口的玻璃管作沸点管，向管内滴加1滴待测液体。另用一根内径约1 mm、长约9 cm的玻璃毛细管作内管，内管一端是封闭的。将内管开口端向下插入沸点管中（见图2-18），用小橡皮圈将沸点管固定在温度计旁，使沸点管底端位于温度计水银球部位，并插入提勒熔点测定管中（见图2-16）。缓缓加热，慢慢升温。不久会观察到有气泡从沸点管内的液体中逸出，这是由于内管中的气体受热膨胀。当升温至液体的沸点时，沸点管中将有一连串的气泡快速逸出。此时，立即停止加热，让浴液自行冷却，管内气体逸出的速度将会减慢。当最后一个气泡因液体的涌入而缩回内管中时，内管内的蒸气压与外界压力正好相等，此时的温度即为该液体在常压下的沸点。

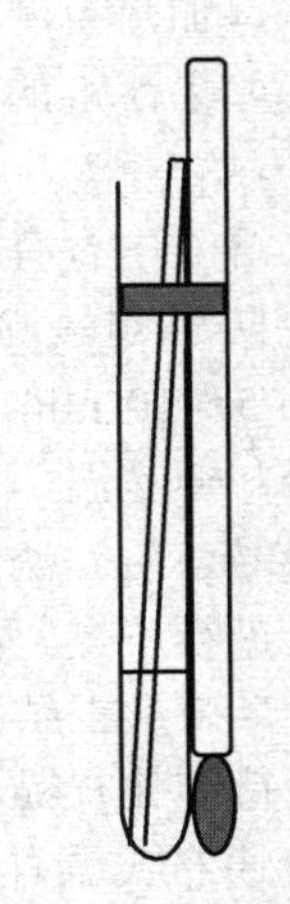

图2-18　沸点测定装置

2.7.3　注意事项

（1）测定沸点时，加热不应过猛，尤其是在接近样品的沸点时，升温更要慢一些，否则沸点管内的液体会迅速挥发而来不及测定。

（2）如果在加热测定沸点过程中，没能观察到一连串小气泡快速逸出，可能是沸点内管封口处没封好之故。此时，应停止加热，换一根内管，待导热液温度降低20℃后即可重新测定。

2.8　重结晶

利用被纯化物质与杂质在同一溶剂中的溶解性能的差异，将其分离的操作称为重结晶（Recrystallization）。重结晶是纯化固体有机化合物最常用的一种方法。

2.8.1　实验原理

固体有机物在溶剂中的溶解度受温度的影响很大。一般来说，升高温度会使溶解度增大，而降低温度则使溶解度减小。如果将固体有机物制成热的饱和溶液，然后使其冷却，这时，由于溶解度下降，原来热的饱和溶液就变成了冷的过饱和溶液，因而有晶体析出。就同一种溶剂而言，对于不同的固体化合物，其溶解性是不同的。重结晶操作就是利用不同物质在溶剂中的不同溶解度，或者经热过滤将溶解性差的杂质滤除；或者让溶解性好的杂质在冷却结晶过程仍保留在母液中，从而达到分离纯化的目的。

2.8.2 实验方法

（1）常量重结晶

对于 1 g 以上的固体样品纯化，一般都采用常量重结晶法。首先将待重结晶的有机物装入圆底烧瓶中，加入少于估算量的溶剂，投入几粒沸石，配置回流冷凝管（见图 2-4）。连通冷凝水，加热至沸，并不时地摇动。如果仍有部分固体没有溶解，再逐次添加溶剂，并保持回流。如果溶剂的沸点较低，当固体全部溶解后再添加一些溶剂，其量约为已加入溶剂量的 15%。

如果溶液中含有色杂质，可以采用活性炭脱色。加入活性炭之前，一定要待上述溶液稍冷却，以防引起暴沸。加入活性炭的量一般为待重结晶有机物投入量的 1%～5%。继续加热，煮沸 5～10 min，用经预热过的布氏漏斗趁热过滤，滤除不溶性杂质和活性炭。所得滤液自然冷却至室温，使晶体析出。然后在室温下过滤，以除去在溶剂中溶解度大的、仍残留在母液中的杂质。滤除母液后，再用少量溶剂对固体收集物洗涤几次，抽干后将晶体放置在表面皿上进行干燥。晶体的纯度可采用熔点测定法进行初步鉴定。

（2）半微量重结晶

如果待纯化样品较少（少于 500 mg）时，用普通布氏漏斗作重结晶操作是比较困难的，一般损失较大，而用 Y 形砂芯漏斗操作，则十分方便，产物损失较小（见图 2-19）。

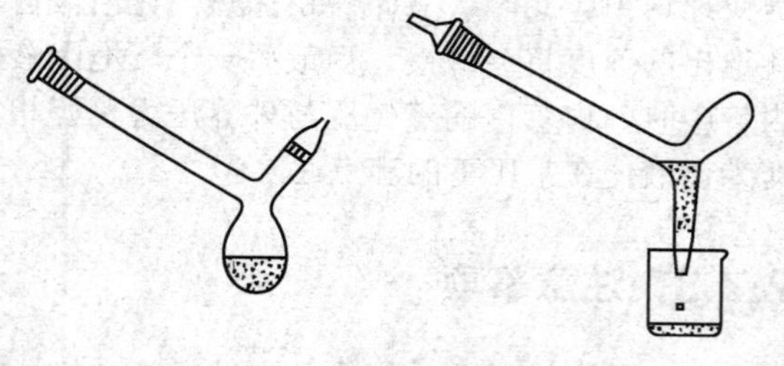

图 2-19　Y 形砂芯漏斗过滤装置

操作时首先将样品由玻璃管口放入球中，加入少许溶剂把落在玻璃管道内的样品冲洗下去，置玻璃球于油浴或热水浴中加热至微沸，再用滴管向球中补加溶剂，直至样品全部溶解。停止热浴，并擦净玻璃球上的油迹或水迹。然后，迅速将玻璃球倒置，用橡皮气球通过玻璃管向 Y 形漏斗内加压，使漏斗内热的饱和溶液经过砂芯漏斗滤入净洁的容器中，静置、结晶。

2.8.3 注意事项

（1）选择适当的溶剂是重结晶过程中一个重要的环节。所选溶剂应该具备以下条件：不与待纯化物质发生化学反应；待纯化物质和杂质在所选溶剂中的溶解度有明显的差异，尤其是待纯化物质在溶剂中的溶解度应随温度的变化有显著的差异；另外，溶剂应容易与重结晶物质分离。如果所选溶剂不仅满足上述条件，而且经济、安全、毒性小、易回收，那就更理想了。

（2）如果所选溶剂是水，则可以不用回流装置。若使用易挥发的有机溶剂，一般都要采用回流装置。

（3）在采用易挥发溶剂时通常要加入过量的溶剂，以免在热过滤操作中，因溶剂迅速挥发导致晶体析出。另外，在添加易燃溶剂时应该注意避开明火。

（4）溶液中若含有色杂质，会使析出的晶体污染；若含树脂状物质更会影响重结晶操作。遇到这种情况，可以用活性炭来处理。通常，活性炭在极性溶液（如水溶液）中的脱色效果较好，而在非极性溶液中的脱色效果要差一些。需要指出的是，活性炭在吸附杂质的同时，对待纯化物质也同样具有吸附作用。因此，在能满足脱色的前提下，活性炭的用量应尽量少。

（5）热过滤操作是重结晶过程中的另一个重要的步骤。热过滤前，应将漏斗事先充分预热。热过滤时操作要迅速，以防止由于温度下降使晶体在漏斗上析出。

（6）热过滤后所得滤液应让其静置冷却结晶。如果滤液中已出现絮状结晶，可以适当加热使其溶解，然后自然冷却，这样可以获得较好的结晶。

（7）经冷却、结晶、过滤后所得的母液，在室温下静置一段时间，还会析出一些晶体，但其纯度就不如第一批晶体。如果对于结晶纯度有一定的要求，前后两批结晶就不可混合在一起。

（8）在用 Y 形管热过滤前，一定要将样品溶液的玻璃球部擦净，否则在倒置过滤时，残留在玻璃球部的溶液可能会污染滤液。

2.9　萃取

用溶剂从固体或液体混合物中提取所需要的物质，这一操作过程就称为萃取（Extraction）。萃取不仅是提取和纯化有机化合物的一种常用方法，而且可以用来洗去混合物中的少量杂质。

2.9.1　实验原理

萃取是利用同一种物质在两种互不相溶的溶剂中具有不同溶解度的性质，将其从一种溶剂转移到另一种溶剂，从而达到分离或提纯目的的一种方法。

在一定温度下，同一种物质（M）在两种互不相溶的溶剂（A，B）中遵循如下分配原理：

$$K=(m_M/V_A)/(m'_M/V'_B)$$

式中，K 表示分配常数；m_M/V_A 表示 M 组分在体积为 V 的溶剂（A）中所溶解的质量（g）；m'_M/V'_B 表示 M 组分在体积为 V' 的溶剂（B）中所溶解的质量（g′）。

换句话说，物质（M）在两种互不相溶的溶剂中的溶解度之比，在一定温度下是一个常数。上式也可以改写为：

$$K=(m_M/m'_M)\times(V'_B/V_A)$$

可见，当两种溶剂的体积相等时，分配常数 K 就等于物质（M）在这两种溶剂中的溶解度之比。显然，如果增加溶剂的体积，溶解在其中的物质（M）量也会增加。

由以上公式还可以推出，若用一定量的溶剂进行萃取，分次萃取比一次萃取的效率高。当然，这并不是说萃取次数越多，效率就越高，一般以提取三次为宜，每次所用萃取剂约相当于被萃取溶液体积的 1/3。

此外，萃取效率还与溶剂的选择密切相关。一般来讲，选择溶剂的基本原则是，对被提取物质溶解度较大；与原溶剂不相混溶；沸点低、毒性小。例如，从水中萃取有机物时常用氯仿、石油醚、乙醚、乙酸乙酯等溶剂，若从有机物中洗除其中的酸或碱或其他水溶性杂质时，可分别用稀碱或稀酸或直接用水洗涤。

以上所述是针对液-液萃取而言。如果要从固体中提取某些组分，则是利用样品中被提取组分和杂质在同一溶剂中具有不同溶解度的性质进行提取和分离的。在实验室中，通常用索氏（Soxhlet）提取器（也称脂肪提取器）从固体中作连续提取操作。其工作原理是通过对溶剂加热回流并利用虹吸现象，使固体物质连续被溶剂所萃取。

2.9.2 实验方法

（1）液-液萃取

将分液漏斗置入固定在铁架台上的铁圈中，把待萃取混合液（体积为 *V*）和萃取剂（体积约为 *V*/3）倒入分液漏斗，盖好上口塞。用右手握住分液漏斗上口，并以右手食指摁住上口塞；左手握住分液漏斗下端的活塞部位，小心振荡，使萃取剂和待萃取混合液充分接触。振荡过程中，要不时将漏斗尾部向上倾斜并打开活塞，以排出因振荡而产生的气体（见图 2-20）。振荡、放气操作重复数次后，将分液漏斗再置放在铁圈上，静置分层。当两相分清后，先打开分液漏斗上口塞，然后打开活塞，使下层液经活塞孔从漏斗下口慢慢放出，上层液自漏斗上口倒出。这样，萃取剂便带着被萃取物质从原混合物中分离出来。一般像这样萃取三次就可以了。将萃取液合并，经干燥后通过蒸馏蒸除萃取剂就可以获得提取物。

（2）液-固萃取

将待提取物研细并用滤纸包起来以细线扎牢，呈圆柱状，置入提取管内。向圆底烧瓶加入溶剂，并投放几粒沸石，配置冷凝管（见图 2-21）。开始加热，使溶剂沸腾，保持回流冷凝液不断滴入提取管中，溶剂逐渐积聚。当其液面高出虹吸管顶端时，浸泡样品的萃取液便会自动流回烧瓶中。溶剂受热后又会被蒸发，溶剂蒸气经冷凝又回流至提取管，如此反复，使萃取物不断地积聚在烧瓶中。当萃取物基本上被提取出来后，蒸除溶剂，即可获得提取物。

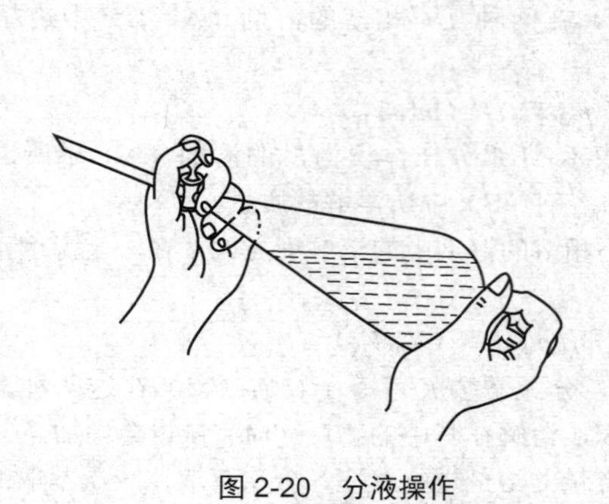

图 2-20　分液操作

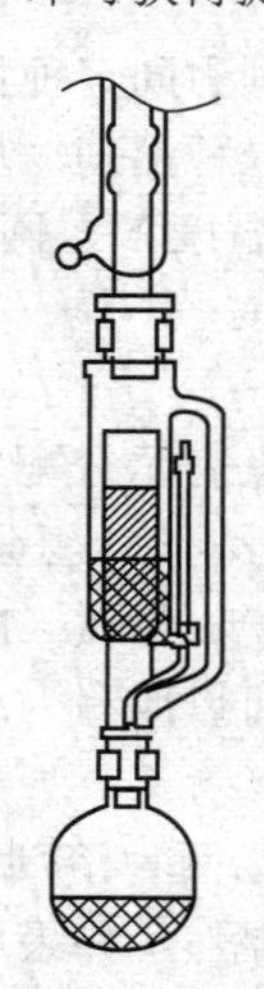

图 2-21　液-固萃取

2.9.3 注意事项

（1）所用分液漏斗的容积一般要比待处理的液体体积大 1～2 倍。在分液漏斗的活塞上应涂上薄薄一层凡士林，注意不要抹在活塞孔中。然后转动活塞使其均匀透明。在萃取操作之前，应先加入适量的水以检查活塞处是否滴漏。

（2）在使用低沸点溶剂（如乙醚）作萃取剂，或使用碳酸钠溶液洗涤含酸液体时，应注意在摇荡过程中要不时地放气。否则，分液漏斗中的液体易从上口塞处喷出。

（3）如果在振荡过程中，液体出现乳化现象，可以通过加入强电解质（如食盐）破乳。

（4）分液时，如果一时不知哪一层是萃取层，则可以通过再加入少量萃取剂来判断：当加入的萃取剂穿过分液漏斗中的上层液溶入下层液，则下层是萃取相；反之，则上层是萃取相。为了避免出现失误，最好将上下两层液体都保留到操作结束。

（5）在分液时，上层液应从漏斗上口倒出，以免萃取层受污染。

（6）如果打开活塞却不见液体从分液漏斗下端流出，首先应检查漏斗上口塞是否打开。如果上口塞已打开，液体仍然放不出，那就该检查活塞孔是否被堵塞。

（7）以索氏提取器来提取物质，最显著的优点是节省溶剂。不过，由于被萃取物要在烧瓶中长时间受热，受热易分解或易变色的物质就不宜采用这种方法。此外，应用索氏提取器来萃取，所使用的溶剂的沸点也不宜过高。

2.10　升华

固体物质受热后不经熔融就直接转变为蒸气，该蒸气经冷凝又可以直接转变为固体，这个过程称为升华（Sublimation）。升华是纯化固体有机物的一种方法。利用升华不仅可以分离具有不同挥发度的固体混合物，而且能除去难挥发的杂质。一般由升华提纯得到的固体有机物纯度都较高。但是，由于该操作较费时，而且损失也较大，因而升华操作通常只限于实验室少量物质的精制。

2.10.1　实验原理

广义地说，无论是由固体物质直接挥发，还是由液体物质蒸发，所产生的蒸气只要是不经过液态而直接转变为固体，这一过程都称为升华。一般来说，能够通过升华操作进行纯化的物质是那些在熔点温度以下具有较高蒸气压的固体物质。这类物质具有三相点，即固、液、气三相并存之点。一种物质的熔点，通常指的是该物质的固、液两相在大气压下达到平衡时的温度。而某物质的三相点指的是该物质在固、液、气三相达到平衡时的温度和压力。在三相点以下，物质只有固、气两相。这时，只要将温度降低到三相点以下，蒸气就可不经液态直接转变为固态。反之，若将温度升高，则固态又会直接转变为气态。由此可见，升华操作应该在三相点温度以下进行。例如，六氯乙烷的三相点温度是186℃，压力为104.0 kPa（780 mmHg），当升温至185℃时，其蒸气压力已达101.3 kPa（760 mmHg），六氯乙烷即可由固相常压下直接挥发为蒸气。

另外，有些物质在三相点时的平衡蒸气压比较低，在常压下进行升华时效果较差，这时可在减压条件下进行升华操作。

2.10.2　实验方法

将待升华物质研细后放置在蒸发皿中，然后用一张扎有许多小孔的滤纸覆盖在蒸发皿口，并用一玻璃漏斗倒置在滤纸上面，在漏斗的颈部塞上一团疏松的棉花（参见图2-22）。

用小火隔着石棉网慢慢加热，使蒸发皿中的物质慢慢升华，蒸气透过滤纸小孔上升，凝结在玻璃漏斗的壁上，在滤纸面上也会结晶出一部分固体。升华完毕，可用不锈钢刮匙将凝结在漏斗壁上以及滤纸上的结晶小心刮落并收集起来。

减压条件下的升华操作与上述常压升华操作大致相同。首先将待升华物质放置在吸滤管内，然后在吸滤管上配置指形冷凝管，内通冷凝水，用油浴加热，吸滤管支口接水泵或油泵（参见图 2-22）。

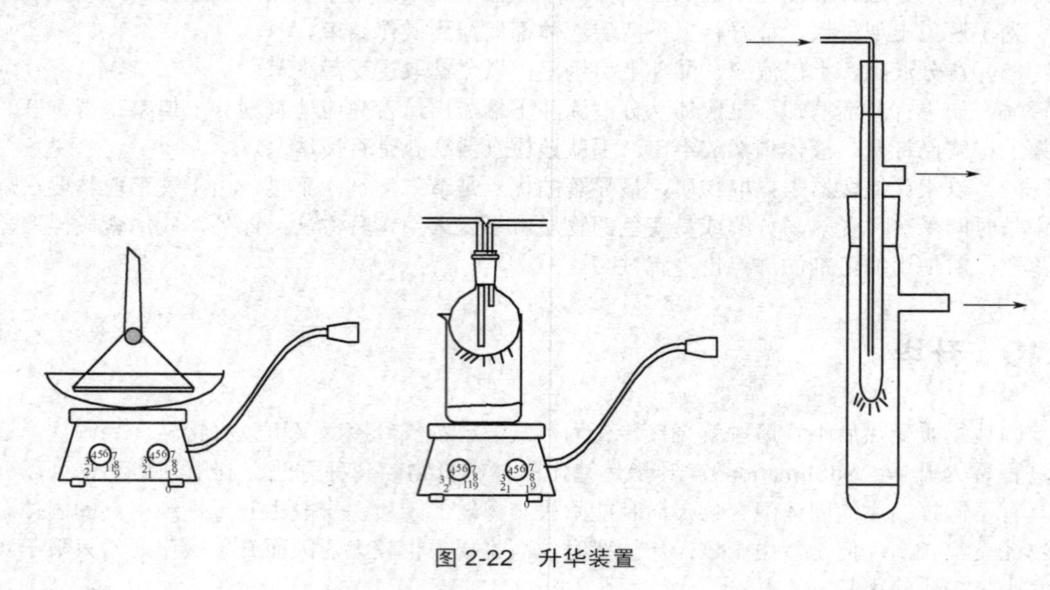

图 2-22　升华装置

2.10.3　注意事项

（1）待升华物质要经充分干燥，否则在升华操作时部分有机物会与水蒸气一起挥发出来，影响分离效果。

（2）在蒸发皿上覆盖一层布满小孔的滤纸，主要是为了在蒸发皿上方形成一温差层，使逸出的蒸气容易凝结在玻璃漏斗壁上，提高物质升华的收率。必要时，可在玻璃漏斗外壁上敷上冷湿布，以助冷凝。

（3）为了达到良好的升华分离效果。最好采取砂浴或油浴而避免用明火直接加热，使加热温度控制在待纯化物质的三相点温度以下。如果加热温度高于三相点温度，就会使不同挥发性的物质一同蒸发，从而降低分离效果。

2.11　色谱法

色谱法（Chromatography）也称色层法或层析法，是分离、提纯和鉴定有机化合物的重要方法之一。色谱法最初源于对有色物质的分离，因而得名。后来，随着各种显色、鉴定技术的引入，其应用范围早已扩展到无色物质。

2.11.1　实验原理

色谱法有许多种类，但基本原理是一致的，即利用待分离混合物中的各组分在某一物质中（此物质称作固定相）的亲和性差异，如吸附性差异、溶解性（或称分配作用）差异等，让混合物溶液（此相称作流动相）流经固定相，使混合物在流动相和固定相之间进行反复吸附或分配等作用，从而使混合物中的各组分得以分离。根据不同的操作条件，色谱法可分为柱色谱（Column chromatography）、纸色谱（Paper chromatography）、薄层色谱（Thin layer chromatography，TLC）。按照流动相的不同，可分为气相色谱（Gas chromatography）、液相色谱（Liquid chromatography）等。

2.11.2 实验方法

（1）柱色谱法

选一合适色谱柱，洗净干燥后垂直固定在铁架台上，色谱柱下端置一吸滤瓶或锥形瓶（参见图 2-23）。如果色谱柱下端没有砂芯横隔，就应取一小团脱脂棉或玻璃棉，用玻璃棒将其推至柱底，然后再铺上一层约 1 cm 厚的砂。关闭色谱柱底端的活塞，向柱内倒入溶剂至柱高的四分之三处。然后将一定量的吸附剂（或支持剂）用溶剂调成糊状，并将其从色谱柱上端向柱内一匙一匙地添加，同时打开色谱柱下端的活塞，使溶剂慢慢流入锥形瓶。在添加吸附剂的过程中，可用木质试管夹或套有橡皮管的玻璃棒轻敲色谱柱，促使吸附剂均匀沉降。添加完毕，在吸附剂上面覆盖约 1 cm 厚的砂层。整个添加过程中，应保持溶剂液面始终高出吸附剂层面（见图 2-23）。

当柱内的溶剂液面降至吸附剂表层时，关闭色谱柱下端的活塞。用滴管将事先准备好的样品溶液滴加到柱内吸附剂表层。用滴管取少量溶剂洗涤色谱柱内壁上沾有的样品溶液。然后打开活塞，使溶剂慢慢流出。当溶液液面降至吸附剂层面时，便可加入洗脱剂进行洗脱。如果被分离各组分有颜色，可以根据色谱柱中出现的色层收集洗脱液；如果各组分无色，先依等分收集法收集，然后用薄层色谱法逐一鉴定，再将相同组分的收集液合并在一起，蒸除溶剂，即得各组分。

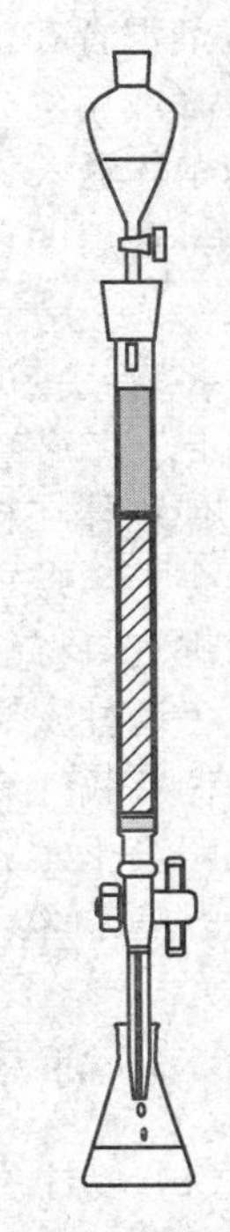

图 2-23　柱色谱

（2）薄层色谱法

将 5 g 硅胶 G 在搅拌下慢慢加入到 12 mL 1%的羧甲基纤维素钠（CMC）水溶液中，调成糊状。然后将糊状浆液倒在洁净的载玻片上，用手轻轻振动，使涂层均匀平整，大约可铺 8 cm×3 cm 载玻片 6～8 块。室温下晾干，然后在 110℃烘箱内活化 0.5 h。

用低沸点溶剂（如乙醚、丙酮或氯仿等）将样品配成 1%左右的溶液，然后用内径小于 1 mm 的毛细管点样。点样前，先用铅笔在色谱板上距末端 1 cm 处轻轻画一横线，然后用毛细管吸取样液在横线上轻轻点样，如果要重新点样，一定要等前一次点样残余的溶剂挥发后再点样，以免点样斑点过大。一般斑点直径不大于 2 mm。如果在同一块色谱板上点两个样，两斑点间距应保持 1～1.5 cm 为宜。

以色谱缸作展开器，加入展开剂，其量以液面高度 0.5 cm 为宜。在展开器中靠瓶壁放入一张滤纸，使器皿内易于达到气液平衡。滤纸全部被溶剂润湿后，将点过样的色谱板斜置于其中，使点样一端朝下，保持点样斑点在展开剂液面之上，盖上盖子（见图 2-24）。当展开剂上升至离色谱板上端约 1 cm 处时，将色谱板取出，并用铅笔标出展开剂的前沿位置。待色谱板干燥后，便可观察斑点的位置。如果斑点无颜色，可将色谱板放置在装有几粒碘晶的广口瓶内盖上瓶盖。当色谱板上出现明显的暗棕色斑点后，即可将其取出，并马上用铅笔标出斑点的位置。然后计算各斑点的 R_f 值。

（3）气相色谱法

选用一根干燥洁净且长度适宜的不锈钢管（有时也可用玻璃管）作色谱柱。根据柱内容积量取比该容积稍多一点的担体，再量取相当于担体质量 5%～25%的固定液，用和担体量相当的低沸点溶剂混合在一起，搅拌均匀。然后利用旋转蒸发仪（或用红外灯加热）蒸除溶剂，将涂有固定液的担体置入 110～120℃的烘箱中老化 2 h。

将选用的色谱柱一端以玻璃毛堵住，并与真空泵相连，另一端连接一个小漏斗。开启真空泵，将老化过的担体逐渐倒入漏斗中，使担体吸入柱内。在装柱过程中，应不断敲击振动色谱柱，使担体在柱中填装得均匀致密。装毕，将漏斗移去，用玻璃毛将色谱柱此端堵住，并以此端作为进气口与色谱仪相连（见图 2-25）。

色谱柱安装在色谱仪的柱箱中，然后开启仪器，调节载气流量（约 10～5 mL/min）和操作温度（略高于实验要求的温度，但低于固定液最高使用温度），待记录仪基线平稳后即可进样测定。

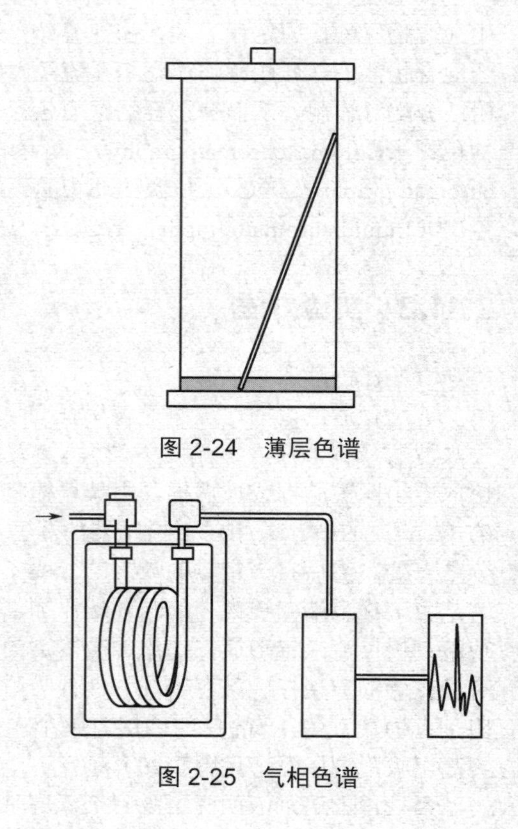

图 2-24　薄层色谱

图 2-25　气相色谱

2.11.3　注意事项

（1）以柱色谱法分离混合物应该考虑到吸附剂的性质、溶剂的极性、柱子的尺寸、吸附剂的用量，以及洗脱的速度等因素。

（2）吸附剂的选择一般要根据待分离的化合物的类型而定。例如酸性氧化铝适合于分离羧酸或氨基酸等酸性化合物；碱性氧化铝适合于分离胺；中性氧化铝则可用于分离中性化合物。硅胶的性能比较温和，属无定形多孔物质，略具酸性，适合于极性较大的物质分离。例如醇、羧酸、酯、酮、胺等。

（3）溶剂的选择一般根据待分离化合物的极性、溶解度等因素而定。有时，使用一种单纯溶剂就能使混合物中各组分分离开来；有时，则需要采用混合溶剂；有时，则使用不同的溶剂交替洗脱。例如，先采用一种非极性溶剂将待分离混合物中的非极性组分从柱中洗脱出来，然后再选用极性溶剂以洗脱具有极性的组分。常用的溶剂有：石油醚、四氧化碳、甲苯、二氯甲烷、氯仿、乙酸、乙酸乙酯、丙酮、乙醇、甲醇、水等。

（4）色谱柱的尺寸以及吸附剂的用量要视待分离样品的量和分离难易程度而定。一般来说，色谱柱的柱长与柱径之比约为 8∶1；吸附剂的用量约为待分离样品质量的 30 倍左右。吸附剂装入柱中以后，色谱柱应留有约四分之一的容量以容纳溶剂。当然，如果样品分离较困难，可以选用更长一些的色谱柱，吸附剂的用量也可适当多一些。

（5）溶剂的流速对柱色谱分离效果具有显著影响。如果溶剂流速较慢，则样品在色谱柱中保留的时间就长，那么各组分在固定相和流动相之间就能得到充分的吸附或分配作用，从而使混合物，尤其是结构、性质相似的组分得以分离。但是，如果混合物在柱中保留的时间太长，则可能由于各组分在溶剂中的扩散速度大于其流出的速度，从而导致色谱带变宽，且相互重叠影响分离效果。因此，洗脱速度要适中。

（6）装柱时要轻轻不断地敲击柱子，以除尽气泡，不留裂缝，否则会影响分离效果。

（7）装柱完毕后，在向柱中添加溶剂时，应沿柱壁缓缓加入，以免将表层吸附剂和样品冲溅泛起，覆盖在吸附剂表层的砂子也是起这个作用。

（8）薄板色谱法除了用于分离提纯外，还可用于有机化合物的鉴定，也可以用于寻找柱色谱分离条件。在有机合成中，还可用来跟踪反应进程。其分离原理是，利用薄层板上的吸附剂在展开剂中所具有的毛细作用，使样品混合物随展开剂向上爬升。由于各组分在吸附剂上受吸附的程度不同，以及在展开剂中溶解度的差异，使其在爬升过程中得到分离。一种化合物在一定色谱条件下，其上升高度与展开剂上升高度之比是一个定值，称为该化合物的比移值，记为 R_f 值。它是用来比较和鉴别不同化合物的重要依据。应该指出，在实际工作中，R_f 值的重现性较差。因此，在鉴定过程中，常将已知物和未知物在同一块薄层板上点样，在相同展开剂中同时展开，通过比较它们的 R_f 值，即可作出判断。

（9）薄板色谱法常用的吸附剂有硅胶和氧化铝，不含黏合剂的硅胶称硅胶 H；掺有黏合剂的硅胶如煅石膏称为硅胶 G；含有荧光物质的硅胶称为硅胶 HF254，可在波长为 254 nm 的紫外光下观察荧光，而附着在光亮的荧光色谱板上的有机化合物却呈暗色斑点，这样就可以观察到那些无色组分；既含煅石膏又含荧光物质的硅胶称为硅胶 GF254。氧化铝也类似地分为氧化铝 G、氧化铝 HF254 及氧化铝 GF254。由于氧化铝的极性较强，对于极性物质具有较强的吸附作用，因而它适合于分离极性较弱的化合物（如烃、醚、卤代烃等）。而硅胶的极性相对较小，它适合于分离极性较大的化合物（如羧酸、醇、胺等）。

（10）制板时，一定要将吸附剂逐渐加入到溶剂中，边加边搅拌。如果颠倒添加次序，把溶剂加到吸附剂中，容易产生结块。

（11）点样时，所用毛细管管口要平整，点样动作要轻快敏捷。否则易使斑点过大，产生拖尾、扩散等现象，影响分离效果。

（12）展开剂的极性差异对混合物的分离有显著影响。当被分离物各组分极性较强，经过色谱后，如果混合物中各组分的斑点全部随溶剂爬升至最前沿，那么该溶剂的极性太强；相反，如果混合物中各组分的斑点完全不随溶剂的展开而移动，则该溶剂的极性太弱。选择展开剂时，可以参考第（3）项。应该指出，有时用单一溶剂不易使混合物分离，这就需要采用混合溶剂作展开剂。这种混合展开剂的极性常介于几种纯溶剂的极性之间。快捷寻找合适的展开剂可以按如下方法操作：先在一块色谱板上点上待分离样品的几个斑点，斑点间留有 1 cm 以上的间距。用滴管将不同溶剂分别点在不同的斑点上，这些斑点将随溶剂向周边扩展形成大小不一的同心圆环。通过观察这些圆环的层次间距，即可大致判断溶剂的适宜性。

（13）碘熏显色法是观察无色物质斑点的一种有效方法。因为碘可以与除烷烃和卤代烃以外的大多数有机物形成有色配合物。不过，由于碘会升华，当色谱板在空气中放置一段时间后，显色斑点就会消失。因此，色谱板经碘熏显色后，应马上用铅笔将显色斑点圈出。如果色谱板上掺有荧光物质，则可直接在紫外灯下观察，化合物会因吸收紫外光而呈黑色斑点。

（14）气相色谱法是以气体作为流动相（即载气）的一种色谱法。根据固定相状态，又分为气-固色谱法和气-液色谱法。实验方法中介绍的是气-液色谱法。气-液色谱法是以多孔惰性固体物质作载体（也称担体），在其表面涂渍一层很薄的高沸点液体有机化合物作为固定相（又称固定液），并将其填充在色谱柱中。当载气将混合物带入色谱柱，混合物各组分将在载气和固定液之间反复进行分配。那些在固定液中溶解度小的组分很快就会被载气带出，而在固定液中溶解度大的组分移动得缓慢，因而各组分被分离开来。气-固色谱法与气-液色谱法原理相似。区别在于气-固色谱法中是以一些多孔固体吸附剂如硅胶、活性氧化铝等直接作固定相。

（15）气相色谱仪型号很多，但它们的组成基本相同，主要包括载气供应系统、进样系统、色谱柱、检测系统以及记录系统等。其操作条件要根据所用机型而定。一般来说，当色谱仪开启稳定后，可用微量注射器进样，气化后的样品经过色谱柱分离成一个个单组分，并依次先后进入检测器，检测器将这些浓度不同的各组分相应地转换为电信号，并以谱峰的形式记录在记录仪上。通常，将从进样开始到往后出现某组分的浓度最大值所需的时间，称保留时间。一般来说，有机化合物在相同的分析条件下，其保留时间是不变的。因此，可以借助气相色谱作定性分析。另外，各组分的含量与其谱峰面积成正比，因而依峰面积大小还可进行定量分析。

（16）在气相色谱操作过程中要用到氢气，切忌明火，注意安全。

2.12 折射率的测定

折射率（Refractive Index）是液体有机化合物的物理常数之一。通过测定折射率可以判断有机化合物的纯度，也可以用来鉴定未知物。

2.12.1 实验原理

在不同介质中，光的传播速度是不相同的，当光从一种介质射入到另一种介质时，其传播方向会发生改变，这就是光的折射现象。根据折射定律，光线自介质 A 射入介质 B，其入射角 α 与折射角 β 的正弦之比和两种介质的折射率成反比。

$$\sin\alpha/\sin\beta = n_B/n_A$$

若设定介质 A 为光疏介质，介质 B 为光密介质，则 $n_A<n_B$。换句话说，折射角 β 必小于入射角 α，见图 2-26。

如果入射角 α=90°，即 $\sin\alpha=1$，则折射角为最大值（称为临界角，以 β_0 示）。折射率的测定都是在空气中进行的，但仍可近似地视作在真空状态之中，即 n_A=1。故有：

$$n=1/\sin\beta_0$$

因此，通过测定临界角 β_0，即可得到介质的折射率 n。通常，折射率是用阿贝（Abbe）折射仪来测定，其工作原理就是光的折射现象。

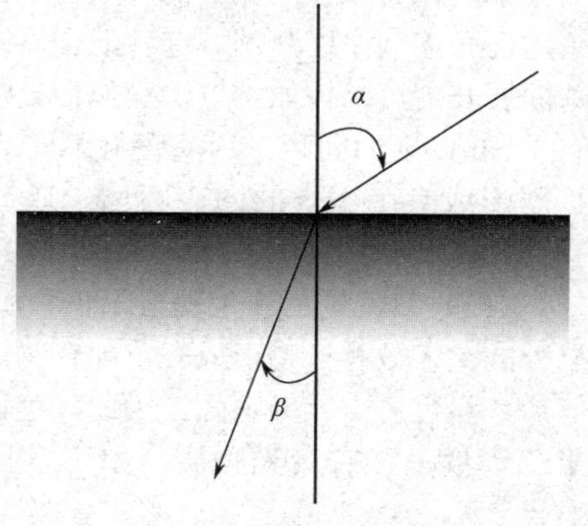

图 2-26 折射现象

由于入射光的波长、测定温度等因素对物质的折射率有显著影响，因而其测定值通常要标注操作条件。例如，在 20℃条件下，以钠光 D 线波长

（589.3 nm）的光线作入射光所测得的四氯化碳的折射率为 1.4600，记为 n_D^{20} 1.4600。由于所测数据可读至小数点后第四位，精确度高，重复性好，因而以折射率作为液态有机物的纯度标准甚至比沸点还要可靠。另外，温度与折射率呈反比关系，通常温度每升高 1℃，折射率将下降 3.5×10^{-4}～5.5×10^{-4}。为了方便起见，在实际工作中常以 4×10^{-4} 近似地作为温度变化常数。例如，甲基叔丁基醚在 25℃时的实测值为 1.3670，其校正值应为：$n_D^{20}=1.3670+5\times4\times10^{-4}=1.3690$

2.12.2　实验方法

打开折射仪的棱镜（见图 2-27），先用镜头纸蘸丙酮擦净棱镜的镜面，然后加 1～2 滴待测样品于棱镜面上，合上棱镜。旋转反光镜，让光线入射至棱镜，使两个镜筒视场明亮。再转动棱镜调节旋钮，直至在目镜中可观察到半明半暗的图案。若出现彩色带，可调节消色散棱镜（棱镜微调旋钮），使明暗界线清晰。接着，再将明暗分界线调至正好与目镜中的十字交叉中心重合（见图 2-28）。记录读数及温度，重复 2 次，取其平均值。测定完毕，打开棱镜，用丙酮擦净镜面。

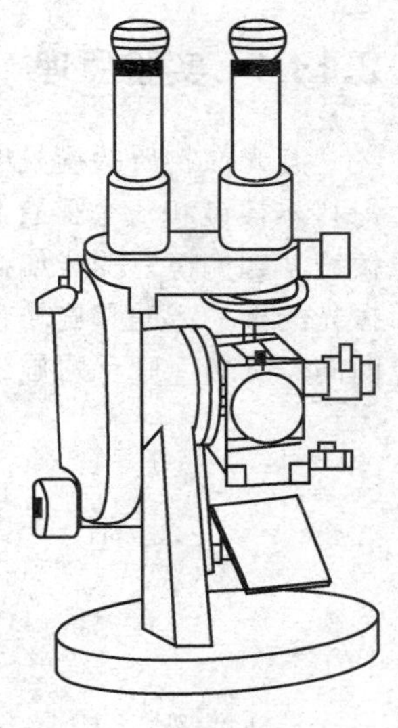

图 2-27　折射仪

2.12.3　注意事项

（1）由于阿贝折射仪设置有消色散棱镜，可使复色光转变为单色光。因此，可直接利用日光测定折射率，所得数据与用钠光时所测得的数据一样。

（2）要注意保护折射仪的棱镜，不可测定强酸或强碱等具腐蚀性液体。

（3）测定之前，一定要用镜头纸蘸少许易挥发性溶剂将棱镜擦净，以免其他残留液影响测定结果。

（4）如果测定易挥发性液体，滴加样品时可由棱镜侧面的小孔加入。

（5）在测定折射率时常见情况如图 2-28 所示，其中图 2-28（4）是读取数据时的图案。当遇到图 2-28（1）即出现色散光带，则需调节棱镜微调旋钮直至彩色光带消失呈图 2-28（2）图案，然后再调节棱镜调节旋钮直至呈图 2-28（4）图案；若遇到图 2-28（3），则是由于样品量不足，需再添加样品，重新测定。

（6）如果读数镜筒内视场不明，应检查小反光镜是否开启。

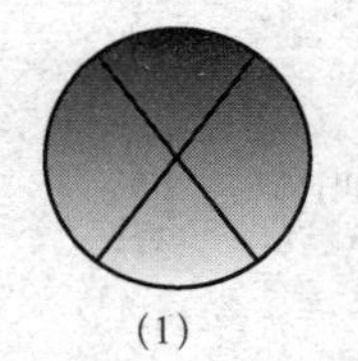

(1)

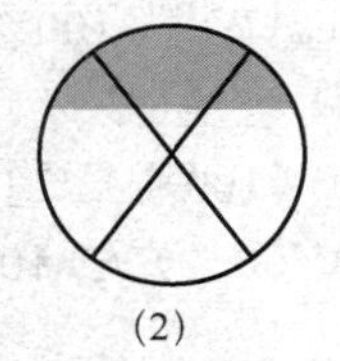

(2)

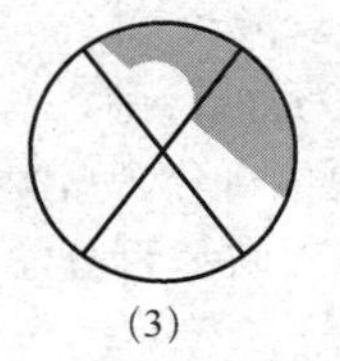

(3)

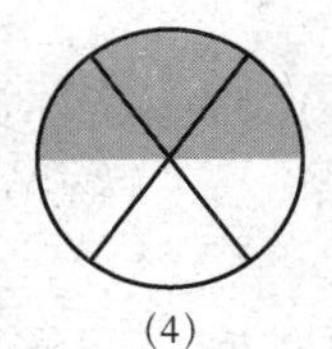

(4)

图 2-28　折射仪中的视野图片

2.13 旋光度的测定

对映体是互为镜像的立体异构体。它们的熔点、沸点、相对密度、折射率以及光谱等物理性质都相同，并且在与非手性试剂作用时，它们的化学性质也一样，唯一能够反映分子结构差异的性质是它们的旋光性不同。当偏振光通过具有光学活性的物质时，其振动方向会发生旋转，所旋转的角度即为旋光度（Optical rotation）。

2.13.1 实验原理

旋光性物质的旋光度和旋光方向可以用旋光仪来测定。旋光仪主要由一个钠光源、两个尼科尔棱镜和一个盛有测试样品的盛液管组成（见图 2-29）。普通光先经过一个固定不动的棱镜（起偏镜）变成偏振光，然后通过盛液管、再由一个可转动的棱镜（检偏镜）来检验偏振光的振动方向和旋转角度。若使偏振光振动平面向右旋转，则称右旋；若使偏振光振动平面向左旋转，则称左旋。

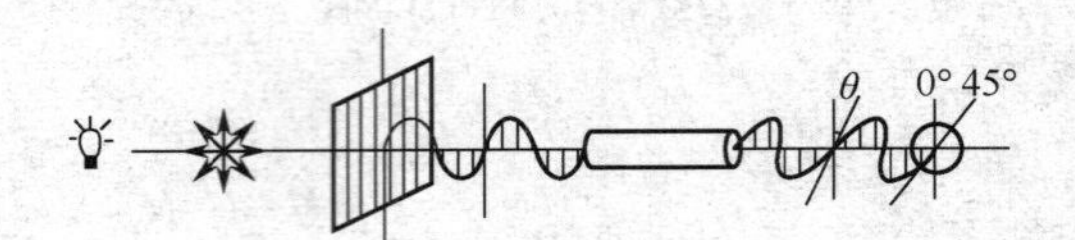

图 2-29　偏振光的生成

光活性物质的旋光度与其浓度、测试温度、光波波长等因素密切相关。但是，在一定条件下，每一种光活性物质的旋光度为一常数，用比旋光度[α]表示：

$$[\alpha]_{\lambda}^{t}=\alpha/(c\cdot l)$$

其中，α 为旋光仪测试值；c 为样品溶液浓度，$g\cdot mL^{-1}$；l 为盛液管长度，dm；λ 为光源波长，通常采用钠光源，以 D 表示；t 为测试温度。如果被测样品为液体，可直接测定而不需配成溶液。求算比旋光度时，只要将其相对密度值（d）代替上式中的浓度值（c）即可：

$$[\alpha]_{\lambda}^{t}=\alpha/(d\cdot l)$$

除了比旋光度外，还可用光学纯度、左旋和右旋对映体的百分含量以及对映体过量值（Enantiomer Excess，e.e）等反映光活性物质的纯度。

若设 s 为旋光异构体混合物中的主要异构体含量，R 为其他异构体含量，则对映体过量 e.e 值用下式计算：

$$e.e.\%=[(s-R)/(s+R)]\times 100$$

若设(−)对映体光学纯度为 $x\%$，则

$$(-)\text{对映体百分含量}\%=[x+(100-x)/2]\times 100\%$$

$$(+)\text{对映体百分含量}=(100-x)/2\times 100\%$$

光学纯度（P）定义为：

$$P=[[\alpha]^{t}_{D}(\text{样品})/[\alpha]^{t}_{D}(\text{标准})]\times 100\%$$

例如，已知样品（S）-(−)-2-甲基丁醇的相对密度 $d_{4}^{23}=0.8$，在 20 cm 长的盛液管中，其旋光为−8.1°，且其标样[α] =−5.8°（纯），则有：

比旋光度$[\alpha]^t_D$（样品）$=\alpha^{23}/(c \cdot l) = -8.1° / (2\times0.8) = -5.1$

光学纯度 $P=$［$[\alpha]^t_D$（样品）/$[\alpha]^t_D$（标准）］$\times100\% = (-5.1)/(-5.8)\times100\% = 88\%$

(−)对映体百分含量=［88+（100−88）/2］×100%=94%

(+)对映体百分含量=（100−88）/2×100%=6%

e.e.%=（$s-R$）/（$s+R$）×100=88%

2.13.2　实验方法

旋光仪有多种类型，现以数字式自动显示旋光仪为例，介绍其操作方法。

（1）预热：打开旋光仪开关，使钠灯加热 15 min，待光源稳定后，再按下“光源”键。

（2）调零：在盛液管中装入用来配制待测样品溶液的溶剂或蒸馏水，将盛液管放置在测试槽中调零，使数字显示屏（或刻度盘）读数为零。

（3）配制溶液：准确称取 0.1～0.5 g 样品，在 25 mL 容量瓶中配成溶液，通常可选用水、乙醇或氯仿作溶剂。若用纯液体样品直接测试，在测试前确定其相对密度即可。

（4）测试：选用适当长度的盛液管，将样品溶液或纯液体样品装入盛液管中，注意除去气泡。然后置盛液管于试样槽中，关上盖。按“测定”键，待数字显示屏（或刻度盘）读数稳定后读数。再重复测量、读数两次，取其平均值。根据公式计算比旋光度、对映体过量值等。

2.13.3　注意事项

（1）如果样品的比旋光度值较小，在配制待测样品溶液时，宜将浓度配得高一些，并选用长一点的测试盛液管，以便观察。

（2）温度变化对旋光度具有一定的影响。若在钠光（λ=589.3 nm）下测试，温度每升高 1℃，多数光活性物质的旋光度会降低 0.3%左右。

（3）测试时，盛液管所放置的位置应固定不变，以消除因距离变化所产生的测试误差。

Fundamental Techniques in Organic Chemistry Experiments

2.1 Common Glass Instruments and Experimental Apparatuses

In organic experiments, glass instruments and experimental apparatuses are often used, and it is necessary to be familiar with these instruments, apparatuses, and their maintenance methods.

2.1.1 Common Glass Instruments

Common glass instruments can be divided into two categories: ordinary glass instruments and standard ground joint instruments.

As the name suggests, standard ground joints have uniform dimensions at the interface, meaning they are standardized. For example, 14/19/24 ground joints refer to the maximum inner diameters of the ground joint, which are 14 mm, 19 mm, and 24 mm, respectively. As long as the ground joints are the same size, they can be assembled together. For ground joint instruments of different sizes, they can be connected by using corresponding sized adapters. When using standard ground joint instruments, alignment during assembly is crucial and excessive force should be avoided to prevent breakage. In general, lubricants are not necessary at the ground joint. However, if performing vacuum distillation, a proper amount of vacuum grease should be applied. After the experiment, the instrument should be disassembled promptly to prevent sticking. Common standard ground joint instruments and other glass instruments are shown in Figure 2-1 to Figure 2-3.

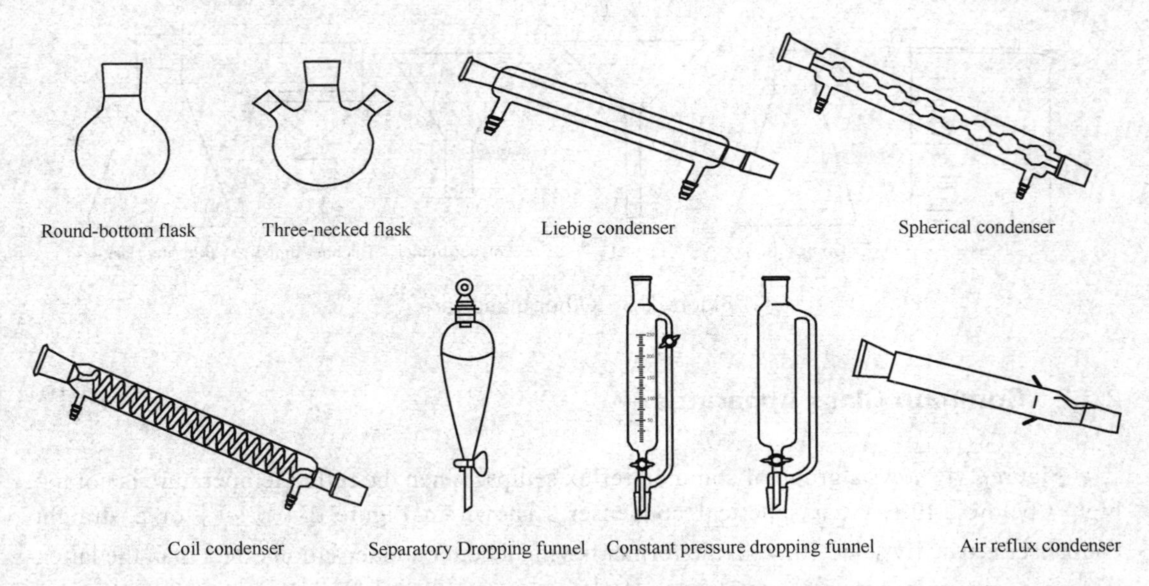

Figure 2-1 Common standard ground glass joint apparatus（Ⅰ）

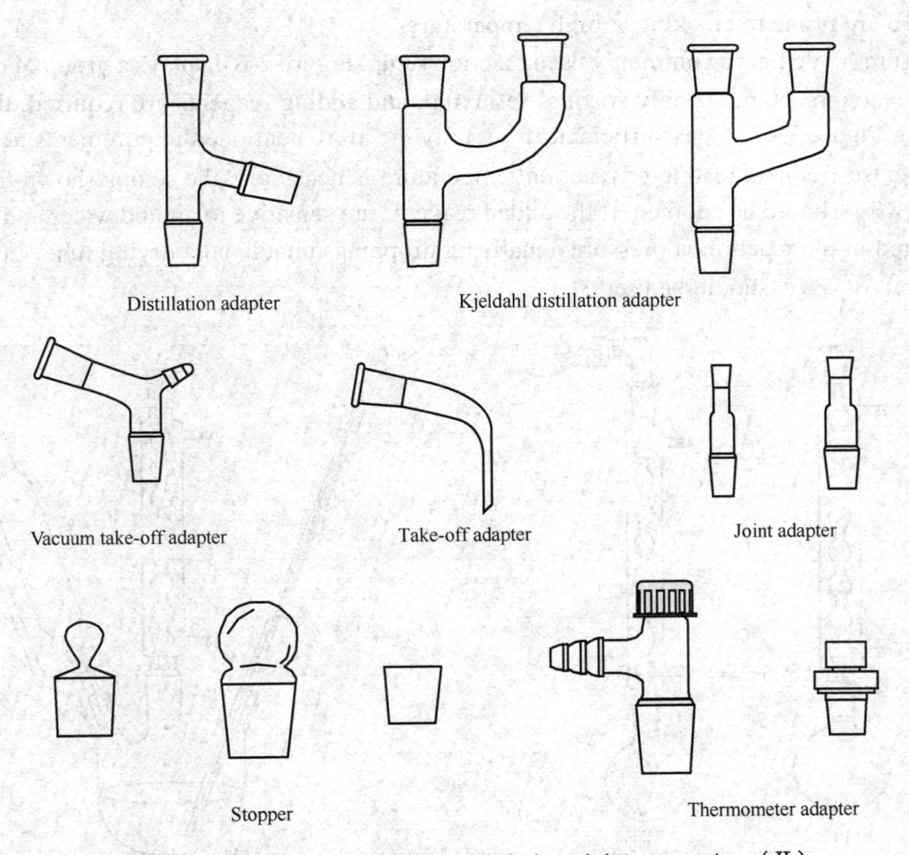

Figure 2-2 Common standard ground glass joint apparatus（Ⅱ）

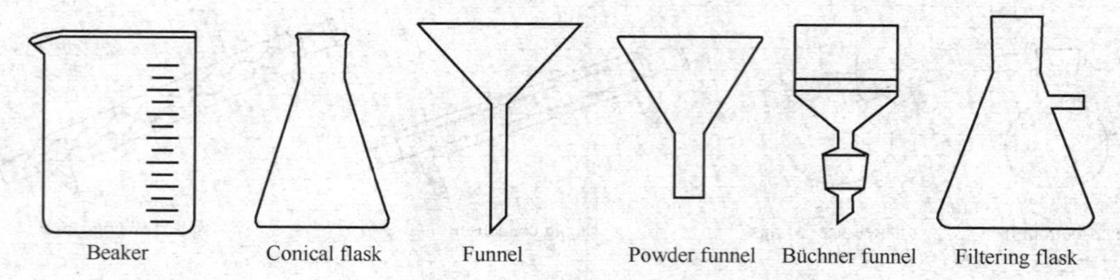

Figure 2-3 Other glassware

2.1.2 Common Glass Apparatus

Figure 2-4 shows a group of common reflux setups. When the reflux temperature is not too high（below 140℃）, a spherical condenser［shown in Figure 2-4（1）］or a straight condenser is usually selected, with the former having a better condensation effect than the latter. If dryness is required in the experiment, a drying tube should also be equipped at the upper end of the condenser ［shown in Figure 2-4（2)］. When the reflux temperature is higher（above 140℃）, an air condenser should be chosen instead of a spherical or straight condenser, as the latter two are prone to cracking at high temperatures.

Figure2-5 depicts a commonly used reaction setup. Figure 2-6 displays a group of common stirring reaction setups. If only stirring, refluxing, and adding reagents are required, the setup shown in Figure 2-6（1）is sufficient. If not only the aforementioned requirements need to be met, but also frequent testing of reaction temperature is necessary, the setups shown in Figure 2-6 and（3）should be adopted. If the added reagents are sensitive to air and water and dryness is required in the reaction, a pressure equalizing dropping funnel and a drying tube（as shown in Figure2-6（3)）should be used.

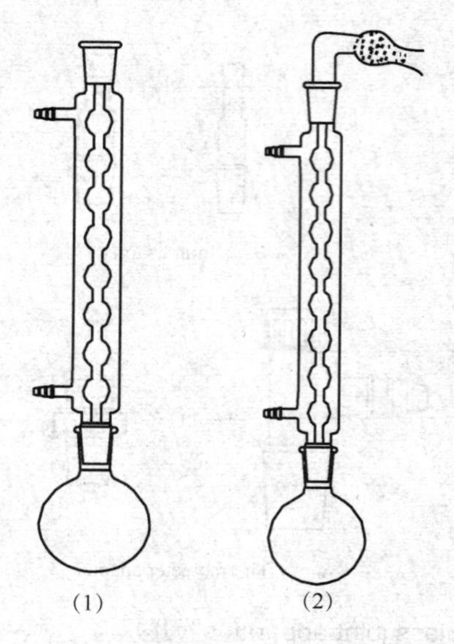

Figure 2-4 Reflux apparatus

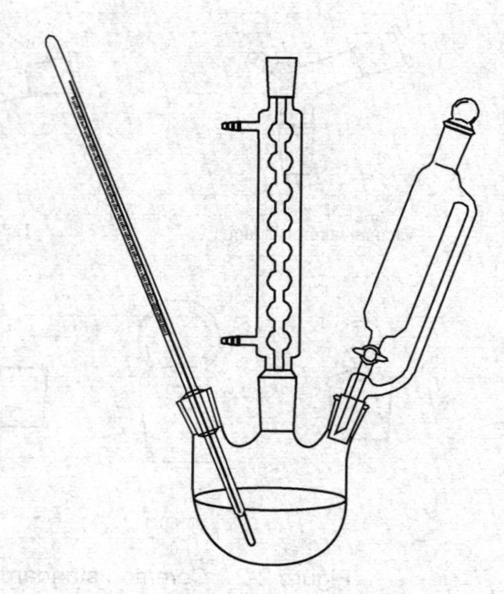

Figure2-5 Reaction apparatus

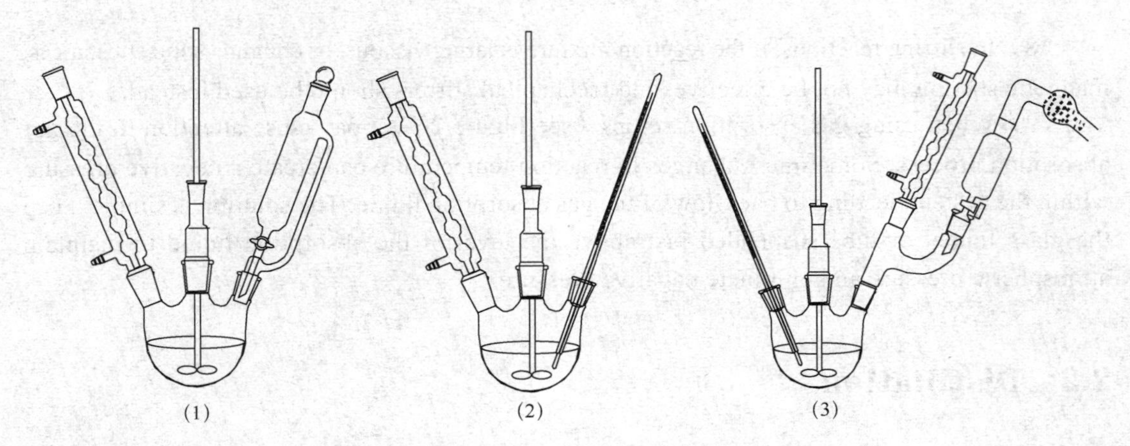

Figure2-6 Stirring Reaction Apparatus

Figure 2-7 presents a group of absorption setups. Gas-absorbing liquid, can be placed in the beaker or filter flask in Figure 2-7 to absorb the alkaline or acidic gas generated during the reaction.

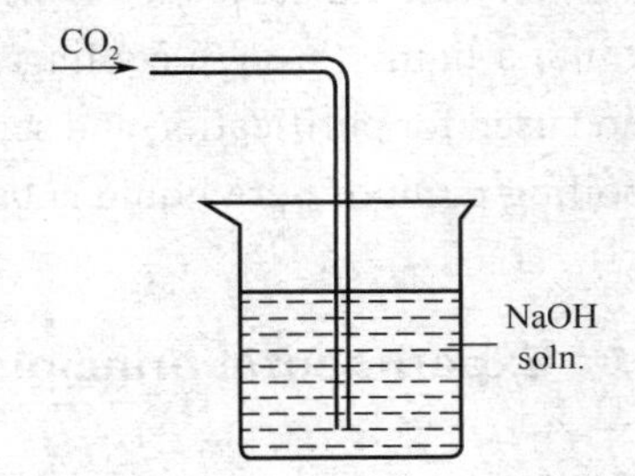

Figure 2-7 Gas Absorption Apparatus

2.1.3 Notes

(1) If the ground glass apparatus is stuck together, do not force it apart. You can use a hairdryer to heat the joint interface where it is stuck and then try to dismantle it.

(2) When heating glassware, except for test tubes, it is generally not advisable to directly heat them with fire to prevent breakage.

(3) Thick-walled glassware such as suction flasks are prone to breakage when heated, so they should not be directly heated. Measuring containers such as graduated cylinders can have their measurement accuracy affected by heat. After cleaning, they should be air-dried instead of being placed at high temperatures for baking.

(4) When glass apparatus with stoppers (such as dropping funnels) are not in use, the stopper and ground glass joints should be separated with paper to avoid sticking.

(5) If alkaline solutions are stored in glassware, they should be promptly washed after use to prevent sticking.

(6) Glassware can be cleaned using detergent or household dishwashing liquid. After cleaning, rinse with water and invert them on a glassware rack to air dry. If quick drying is needed, a small amount of acetone or ethanol can be used for rinsing, followed by drying with a hairdryer.

(7) In reflux setups, spherical condensers are commonly used. Due to the larger contact area between vapor and the condenser, they provide better condensation effects, especially for low-boiling solvents. However, when the reflux temperature exceeds 140℃, an air condenser should be used.

(8) In stirring reactions, if the reaction mixture is large, viscous, or contains solid substances, magnetic stirring may not be effective, and mechanical stirrers should be used instead.

(9) When using gas absorption setups (see Figure 2-7), pay close attention to the gas absorption process. Sometimes, changes in reaction temperature can create a negative pressure within the system, leading to backflow of the gas absorption liquid. The solution is simple: keep the glass funnel or tube suspended just above the level of the absorption liquid to maintain atmospheric pressure and eliminate negative pressure.

2.2 Distillation

When a liquid substance is heated and converted into vapor, the vapor can be condensed back into a liquid through cooling. This process is called distillation. Distillation is a common method used for purification and separation of liquid substances. It can also be used to determine the boiling point of pure liquid substances.

2.2.1 Experimental principle

Pure liquid substances have a specific boiling point under certain pressure, and different substances have different boiling points. Distillation takes advantage of the difference in boiling points of different substances to separate and purify liquid mixtures. When a liquid mixture is heated, the component with the lower boiling point evaporates first and is distilled out, while the component with the higher boiling point either does not evaporate easily or its vapors condense and remain in the distillation flask, thus separating the mixture. However, for distillation to be effective, the difference in boiling points between the components should be greater than 30℃. If the difference is small, fractional distillation is required to separate and purify the liquid mixture.

It should be noted that liquids with constant boiling points are not necessarily pure compounds, as some compounds can form binary or ternary azeotropic mixtures, which cannot be separated by distillation. Pure compounds usually have a narrow boiling range (about 0.5 to 1℃), while mixtures have a wider boiling range. Therefore, distillation can be used not only for qualitative identification of compounds but also for determining their purity.

2.2.2 Experimental procedure

Set up the distillation flask, condenser, receiver, and delivery tube (see Figure 2-8). Pour the liquid to be distilled into the flask, add 1～2 pieces of boiling stones, and insert a thermometer.

Turn on the cooling water and start heating to make the liquid in the flask boil. Adjust the flame and control the distillation rate to about 1～2 drops per second. During the distillation process, observe the temperature reading on the thermometer and record the temperature when the first drop of distillate is collected. Once the temperature reading stabilizes, switch to a new receiver to collect the fractions. If heating is continued but no more distillate is obtained and the temperature suddenly decreases, it indicates that the fraction in that range has been distilled, and heating should be stopped. Record the boiling range and volume (or mass) of that fraction. The smaller the temperature range of the fraction, the higher its purity.

Sometimes, after an organic reaction, it is necessary to directly distill the reaction mixture. In such cases, a three-necked flask can be assembled as the distillation apparatus for direct distillation (see Figure 2-9).

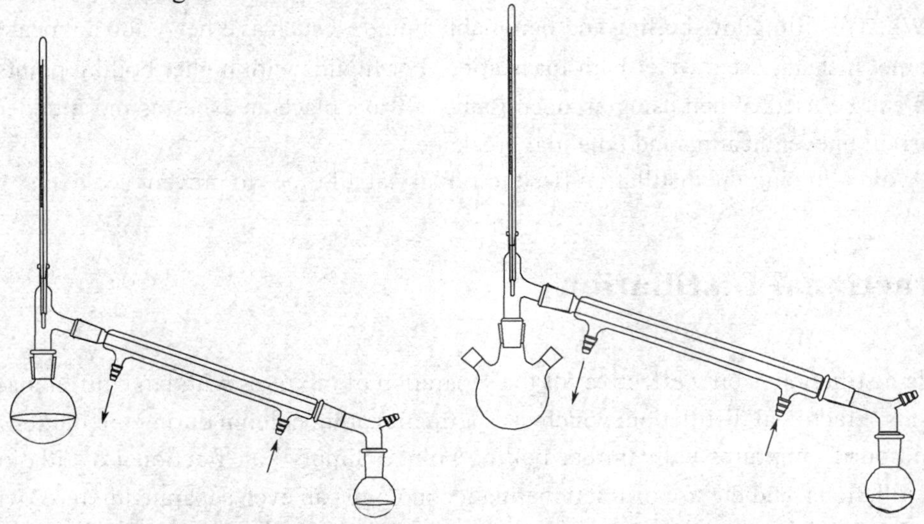

Figure 2-8 Simple distillation apparatus diagram

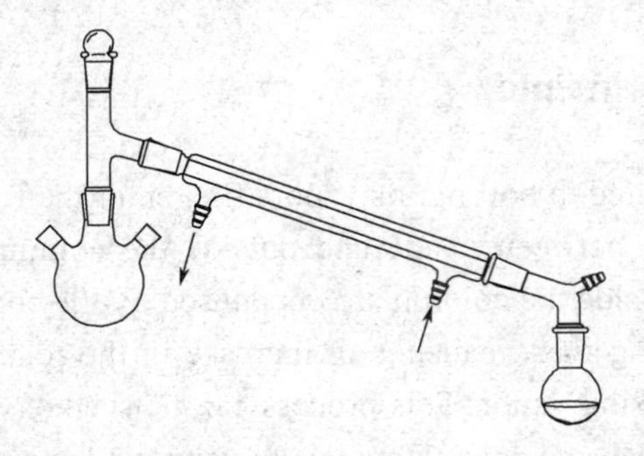

Figure 2-9 Distillation apparatus modified from a reaction setup

2.2.3 Notes

(1) The size of the distillation flask should be chosen according to the volume of the liquid to be distilled. Typically, the volume of the liquid should occupy about 1/3 to 2/3 of the flask.

(2) When the boiling point of the liquid to be distilled is below 140℃, a straight condenser should be used. If the boiling point is above 140℃, an air condenser should be used. Using a straight condenser, in this case, may cause breakage.

(3) If there is no sidearm on the delivery tube used in the distillation setup, there should be a gap between the delivery tube and the receiver to ensure that the system is open to the atmosphere. Otherwise, accidents may occur when the closed system is heated.

(4) Boiling stones are porous materials such as porcelain chips or capillary tubes. When the liquid is heated and starts to boil, the small bubbles in the boiling stones act as nucleation sites, ensuring smooth boiling of the liquid. If distillation has already started and you realize that you forgot to add boiling stones, do not add them directly to the hot liquid to avoid a sudden and violent boiling. Instead, stop heating, let the liquid cool slightly, and then add the boiling stones.

(5) When distilling low-boiling and flammable liquids (such as ether), do not heat with an open flame. Instead, use a water bath for heating. For liquids with higher boiling points, an open flame can be used. When using an open flame, always place an asbestos mesh under the flask to prevent uneven heating and potential breakage.

(6) Avoid allowing the distillation flask to run dry at all times to prevent accidents.

2.3 Fractional Distillation

Simple distillation is only effective for the separation of mixtures with large differences in boiling points. Fractional distillation, which uses a fractionating column during distillation, can separate and purify mixtures with similar boiling points. Simply put, fractional distillation is multiple distillations, and the use of fractionating technology can even separate mixtures with a difference of 1-2°C in boiling point.

2.3.1 Experimental principle

When a mixture is heated to boiling, its vapors first enter the fractionating column. Due to the temperature difference between inside and outside the column, the higher boiling point components in the vapor inside the column are condensed by the cooler air outside the column, and flow back to the boiling flask, causing an increase in the relative amount of low boiling point components in the rising vapor. This process can be seen as a simple distillation. When the high boiling point condensate encounters newly vaporized vapor on the way back up, heat exchange occurs between the two, and high boiling point components in the rising vapor are also condensed, while the low boiling point components continue to rise. This can also be seen as a simple distillation. Vapor repeatedly undergoes the processes of vaporization, condensation, and reflux in the fractionating column, or in other words, multiple simple distillations. Therefore, as long as the efficiency of the fractionating column is high enough, the vapor components distilled from the upper end of the column can approach the purity of low boiling point single components, while the high boiling point components continue to reflux into the distillation flask. It should be noted that, like distillation, fractional distillation cannot be used to separate azeotropic mixtures.

2.3.2 Experimental procedure

Place the substance to be fractionated into a round-bottom flask and add a few beads of zeolite. Then, install the fractionating column, thermometer, condenser, receiving flask, and transfer tube in sequence（see Figure 2-10）.

Turn on the cooling water and begin heating until the liquid boils steadily. When the vapor rises slowly, control the temperature to maintain a distillation rate of 2～3 seconds per drop. Record the temperature when the first drop of distillate drips into the receiving flask, then collect the fractions in sections according to specific requirements, and record the boiling point range and volume of each fraction.

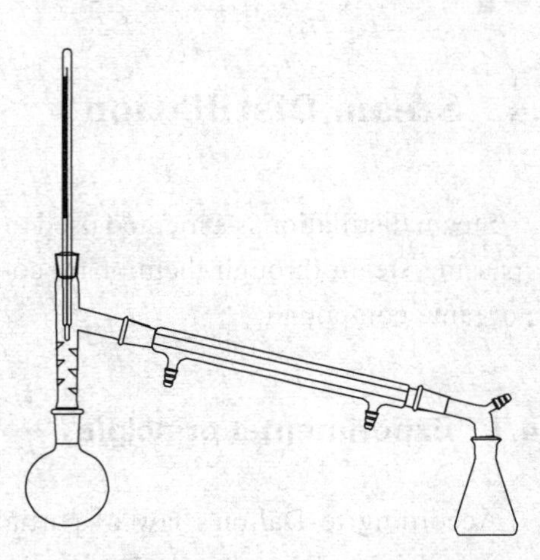

Figure 2-10　Fractional distillation apparatus diagram

2.3.3 Notes

（1）The height of the fractionating column is an important factor affecting the separation efficiency. Generally speaking, the higher the fractionating column, the more times there are heat exchanges between the rising vapor and the condensate, and the better the separation effect. However, if the fractionating column is too high, it will affect the distillation rate.

（2）The filling material inside the fractionating column is also an important factor affecting the separation efficiency. The filling material increases the contact between the vapor and the reflux liquid in the column. The larger the surface area to volume ratio of the filling material, the more favorable it is to improve the separation efficiency. However, it should be noted that there should be a certain gap between the filling materials, otherwise it will cause difficulties in distillation. The Vigreux column is a simple fractionating column with a thorn-like structure inside and does not require additional filling material, which is commonly used in laboratories.

（3）When the room temperature is low or the boiling point of the substance to be fractionated is high, the insulation performance of the fractionating column will significantly affect the separation efficiency. In this case, if the insulation performance of the fractionating column is poor, heat dissipation will be fast, making it difficult to maintain thermal balance between the gas-liquid phases inside the column, thereby affecting the separation effect. To improve the insulation performance of the fractionating column, glass cloth and other insulation materials can be used to wrap the column.

（4）In the process of fractional distillation, attention should be paid to adjusting the heating temperature to maintain an appropriate distillation rate. If the distillation rate is too fast, liquid flooding may occur, that is, the reflux liquid cannot flow back to the boiling flask in time and gradually forms a liquid column in the fractionating column. If this phenomenon occurs, stop heating, wait for the liquid column to disappear, and then resume heating to reach a gas-liquid equilibrium before collecting the fractions.

2.4 Steam Distillation

Steam distillation is a method used to separate and purify liquid or solid organic compounds by passing steam through them or by co-distilling them with water, which is immiscible with the organic compound.

2.4.1 Experimental principle

According to Dalton's law of partial pressures, when water and organic compounds are mixed and heated together, the vapor pressure is the sum of the vapor pressures of each component. In other words:

$$p_{\text{mixture}} = p_{\text{water}} + p_{\text{organic}}$$

If the sum of the vapor pressures of water and organic compounds is equal to the atmospheric pressure, the mixture will boil and both the organic compounds and water will be vaporized together. Obviously, the boiling temperature of the mixture will be lower than the boiling point of either component. In other words, organic compounds can be vaporized at temperatures below their boiling points. Theoretically, the ratio of the weight of organic compounds (m_{organic}) to water (m_{water}) in the distilled liquid should be equal to the ratio of their respective vapor pressures (p_{organic} and p_{water}) multiplied by their respective molecular weights (M_{organic} and M_{water}).

For example, when 1-octanol is subjected to steam distillation, the mixture of 1-octanol and water boils at 99.4°C. By consulting the handbook, it is easy to find that pure water has a vapor pressure of 99.18 kPa (744 mmHg) at 99.4℃. According to Dalton's law of partial pressures, the sum of the vapor pressure of water and 1-octanol is 101.31 kPa (760 mmHg). Therefore, the vapor pressure of 1-octanol at 99.4°C is 2.13 kPa (16 mmHg). Thus, for every gram of water vaporized, 0.16 grams of 1-octanol is vaporized.

Since the temperature of co-boiling of organic compounds and water is always below 100℃, steam distillation is particularly suitable for the separation of organic compounds that undergo changes at high temperatures. Of course, the organic compound must also have a vapor pressure of at least 0.7 kPa (5 mmHg) and be immiscible with water. In addition, mixtures containing a large amount of resinous impurities or mixtures that are difficult to separate by direct distillation or recrystallization can also be separated by steam distillation.

2.4.2 Experimental procedure

Install the steam generator, round-bottom flask, Claisen adapter, thermometer, condenser, receiver, and receiving flask in sequence [see Figure 2-11 (1)]. Transfer the mixture to be separated into the flask, open the side arm of the Claisen adapter, heat the steam generator to boil the water. When steam sprays out from the side arm, close the side arm and allow the steam to enter the flask. Connect the cooling water to quickly condense the mixed vapor in the condenser and collect it in the receiving flask. The distillation rate should be around 2 drops per second, controlled by adjusting the flame. When the distillate becomes clear and transparent and no longer contains oily substances, stop the distillation. First, open the side arm of the Claisen adapter, then stop heating. Transfer the collected liquid to a separatory funnel, allow it to settle into layers, remove the water layer, and obtain the separated product. If a simpler steam distillation apparatus is used instead of a steam generator, the steam distillation operation can still be carried out normally [see Figure2-11 (2)]. First, transfer the mixture to be separated, along with a suitable amount of water, into the round-bottom flask. Add a few pieces of boiling stones, connect the condenser and start heating to maintain a steady boil. The other steps are the same as described earlier, except that when the water in the flask decreases due to continuous distillation, additional water can be added through the dropping funnel on top of the Claisen adapter. It's easy for the mixture to splash into the condenser if using the apparatus shown in Figure2-11 (2), affecting the separation and purification, then using Figure2-11 (3) can effectively avoid this problem. However, due to the long bent section of the Claisen adapter, the vapor easily condenses, which affects efficient distillation. In this case, insulating materials such as glass wool can be wrapped around to prevent rapid heat loss and improve distillation efficiency.

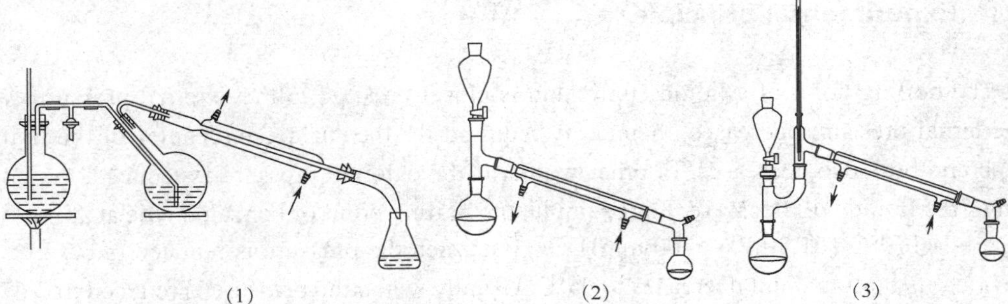

Figure 2-11 Steam Distillation

2.4.3 Notes

(1) A safety tube must be installed in the steam generator. A long glass tube can be used as the safety tube, with its lower end close to the bottom of the steam generator. When in use, do not fill the steam generator with too much water, generally not exceed 2/3 of its capacity.

(2) The connecting pipe between the steam generator and the round-bottom flask should be as short as possible to minimize heat loss during the introduction of steam.

(3) The glass tube for introducing steam should be placed as close to the bottom of the round-bottom flask as possible to improve distillation efficiency.

(4) During the distillation process, if a significant amount of steam condenses and

accumulates in the round-bottom flask, the bottom of the flask can be heated gently over a low flame with an asbestos net.

(5) During the experiment, regularly observe the safety tube. If the water column shows abnormal rising, immediately open the side arm of the Claisen adapter, stop heating, identify the cause, troubleshoot, and then restart the distillation.

(6) When stopping the distillation, it is necessary to open the side arm of the Claisen adapter first, then stop heating. If the heating is stopped first, the steam generator will cool down and create a negative pressure, resulting in back-suction of the mixture in the flask.

2.5 Vacuum Distillation

Some organic compounds have poor thermal stability and often undergo decomposition, oxidation, or polymerization before reaching their boiling point at the heating temperature. For the purification or separation of these compounds, it is not suitable to use atmospheric distillation, but rather distillation under reduced pressure. Vacuum distillation, also known as vacuum distillation, can distill organic compounds at temperatures below their boiling points. Vacuum distillation is especially suitable for distilling organic compounds with high boiling points and poor thermal stability.

2.5.1 Experimental principle

The boiling point of a liquid compound is closely related to the external pressure. When the external pressure decreases, the energy required for the surface molecules of the liquid to escape and boil also decreases. In other words, if the external pressure is reduced, the boiling point of the liquid will decrease. For example, the boiling point of benzaldehyde at atmospheric pressure is 179°C/101.3 kPa (760 mmHg), but when the pressure is reduced to 6.7 kPa (50 mmHg), its boiling point decreases to 95°C. Usually, when the pressure is reduced to 2.67 kPa (20 mmHg), the boiling point of most organic compounds is about 100℃ lower than their boiling point at atmospheric pressure. The relationship between boiling point and pressure can be approximately derived from Figure 2-12. For example, if the boiling point of a compound at atmospheric pressure is 200°C, and you want to distill it under a reduced pressure of 4.0 kPa (30 mmHg), what is the boiling point? First, find the 200°C mark on the atmospheric boiling point scale line in Figure 2-12, then find the 4.0 kPa (30 mmHg) mark on the system pressure curve. Connect these two points with a straight line and extend them to intersect with the reduced pressure boiling point scale line. The number indicated by the intersection point is the boiling point of the compound under the reduced pressure of 4.0 kPa (30 mmHg), which is 100°C. In the absence of other sources of information, the estimated value obtained by this method still has a certain reference value for actual vacuum distillation operations.

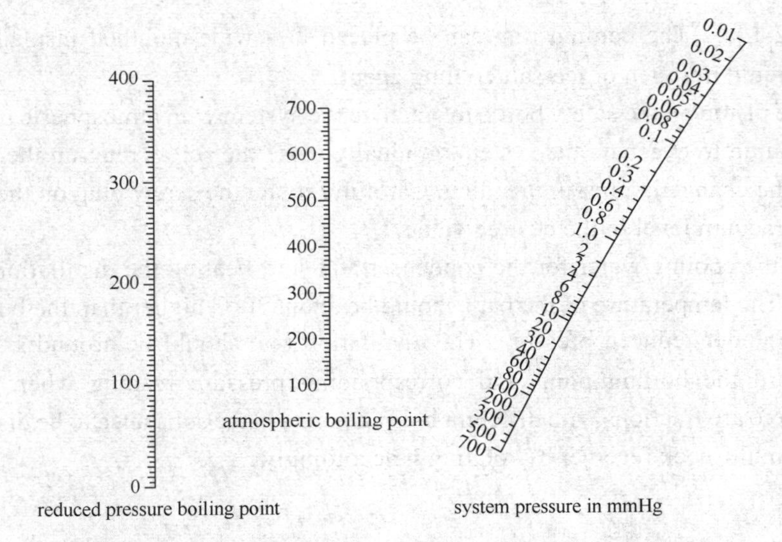

Figure 2-12 Approximate relationship between the boiling points of liquids under atmospheric and reduced pressure

2.5.2 Experimental procedure

Usually, a vacuum distillation system consists of a distillation apparatus, safety bottle, gas absorption apparatus, buffer bottle, and pressure measurement device. When performing vacuum distillation, assemble the distillation flask, Kjeldahl distillation, condenser, vacuum adapter, and receiving flask in order. Use a glass funnel to introduce the substance to be distilled into the distillation flask, with the capillary tube positioned as close to the bottom of the flask as possible [see Figure 2-13 (1)] .

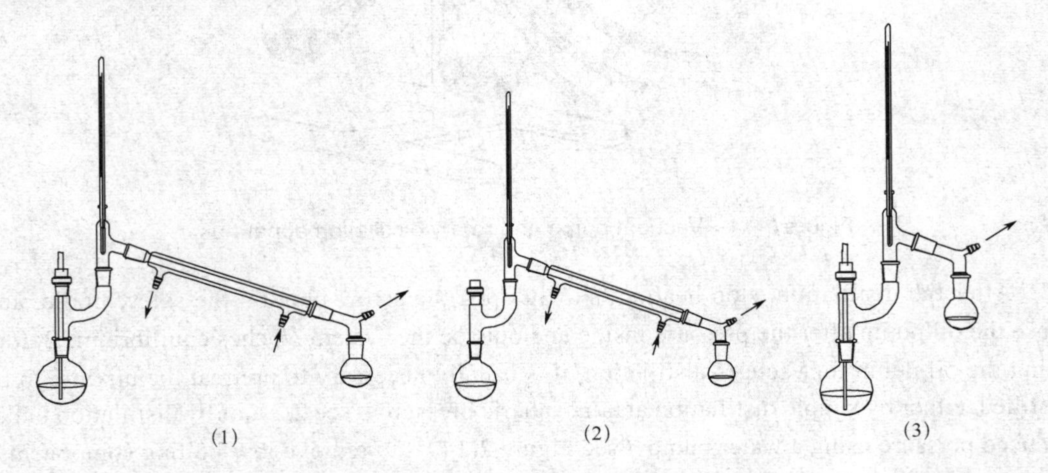

Figure 2-13 Vacuum distillation apparatus

Connect the vacuum adapter sequentially to the safety bottle, cooling trap, vacuum gauge, gas absorption tower, buffer bottle, and oil pump using a thick-walled vacuum rubber tube (see Figure 2-14) . The cooling trap can be placed in a wide-mouthed insulated bottle and cooled with liquid nitrogen or ice-salt cooling agent.

Open the piston on the safety bottle to equalize the system with atmospheric pressure. Then start the oil pump to evacuate the system, gradually close the screw plug on the safety bottle, and observe the change in pressure reading. Carefully rotate the screw plug on the safety bottle to adjust the vacuum level to the desired value.

Turn on the cooling water for the condenser and start heating the distillation flask with a heating bath. The temperature of the bath should be about 30℃ higher than the boiling point of the substance under reduced pressure. The distillation rate should be around 1～2 drops per second. Record the boiling point and corresponding pressure reading when fractions are distilled. If there are fractions with different boiling points in the substance to be distilled, collect them using a multi-neck receiver by rotating it accordingly.

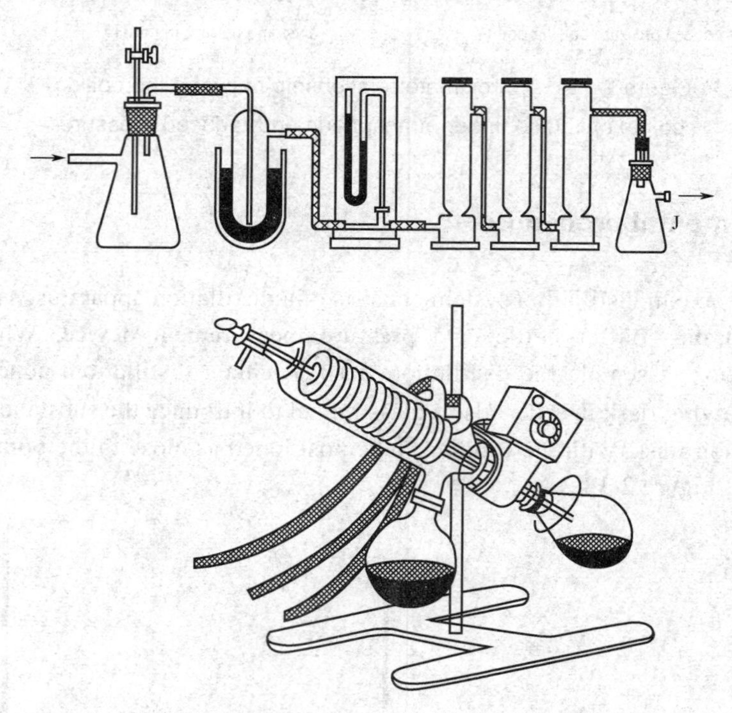

Figure 2-14 Vacuum pump and rotary distillation apparatus

After the distillation, stop heating, slowly open the screw plug on the safety bottle, and close the oil pump after the pressure inside and outside the system reaches equilibrium. Before using the oil pump for vacuum distillation, it is usually necessary to pretreat the mixture to be distilled, either by simple distillation at atmospheric pressure (see 2.2) or by distillation under reduced pressure using a water pump (see Figure 2-14) , to remove low-boiling components.

2.5.3 Notes

(1) In the vacuum distillation setup, there is a capillary tube directly inserted into the

bottom of the distillation flask from the kjeldahl distillation. It serves to introduce the vaporization center and ensure smooth distillation. If a magnetic stirrer is installed in the distillation flask, turning on the magnetic stirrer during vacuum distillation can also maintain stable distillation, and there is no need to install the capillary tube [see Figure 2-13 (2)]. If the substance to be distilled is sensitive to air, vacuum distillation with magnetic stirring is more suitable. In this case, if a capillary tube is still used, it should be protected by introducing inert gas (such as nitrogen) through the capillary tube.

(2) After turning on the oil pump, pay attention to the pressure gauge. If there is no significant change in system pressure or if the system fails to reach the desired vacuum level, check for leaks in the system. Before checking, turn off the oil pump and then inspect the connecting parts section by section. If there is a leak in the distillation apparatus, apply a small amount of vacuum grease to the appropriate joints and rotate them for a tight seal. If there is a leak in other connected parts, such as the gas absorption tower and pressure gauge, apply a small amount of melted paraffin and heat it with a hairdryer (or apply vacuum grease). After the inspection is completed, the oil pump can be turned on according to the procedure described in the experimental method.

(3) During vacuum distillation, uniform heating should be achieved using an oil bath (or water bath). Generally, the bath temperature should be about 30°C higher than the boiling point of the substance under reduced pressure.

(4) If distilling a small amount of high-boiling substances or low-melting substances, distillation can be performed using the setup shown in Figure 2-13 (3), omitting the condenser. If the distillation temperature is high, to reduce heat loss, insulation materials such as glass wool can be wrapped around the Kjeldahl distillation. If the boiling point of the liquid is below 140～150°C under reduced pressure, the receiving flask can be cooled with a cold-water bath.

(5) When using the oil pump, attention should be paid to protection and maintenance, and moisture, organic substances, or acidic gases should not enter the pump, as it will severely reduce the efficiency of the oil pump. The safety bottle, cooling trap, gas absorption tower, and buffer bottle installed between the distillation apparatus and the oil pump are designed to protect the oil pump. If a sudden bumping or surge occurs during distillation, the safety bottle serves as a protective device. Sometimes, due to a sudden change in system pressure, causing oil backflow, the buffer bottle can prevent the oil from entering the gas absorption tower. In addition, for components with lower boiling points that are drawn out, the cooling trap can be immersed in a wide-mouthed insulated bottle containing liquid nitrogen, dry ice, ice-water, or ice-salt cooling agents, depending on the specific situation. The gas absorption tower, also known as the drying tower, is generally equipped with 2-3 units. These drying towers are filled with anhydrous calcium chloride, granular sodium hydroxide, and solid paraffin to absorb moisture, acidic gases, and hydrocarbon gases. It should be noted that before using the oil pump for vacuum distillation, simple distillation or distillation under reduced pressure using a water pump should be performed first to remove low-boiling substances and prevent them from being sucked into the oil pump.

(6) Figure 2-15 shows a closed-type mercury pressure gauge commonly used to measure

the degree of vacuum in a vacuum system. The difference in the mercury levels in the two arms indicates the vacuum level of the vacuum system. During use, when the vacuum operation is complete, slowly open the double-ended screw plug on the safety bottle, allowing gas to enter the system gradually and causing the mercury column in the pressure gauge to slowly return to its original position, in order to prevent the sudden increase in pressure in the system from causing the mercury column to break through the glass tube.

Figure 2-15 closed-type mercury pressure gauge

2.6 Melting Point Determination

The temperature at which a compound undergoes a solid-to-liquid transition under atmospheric pressure is called its melting point. Melting point is one of the physical constants of solid organic compounds, and it can be used not only to identify different organic compounds but also to determine their purity.

2.6.1 Experimental principle

Strictly speaking, the melting point refers to the temperature at which a compound reaches equilibrium between its solid and liquid phases under atmospheric pressure. Typically, pure organic compounds have well-defined melting points, and the temperature range from initial melting to complete melting (referred to as melting range or melting interval) is narrow, usually not exceeding 0.5～1°C. However, if impurities are present in the sample, it can cause a decrease in the melting point and a broadening of the melting range. Therefore, by measuring the melting point and observing the melting range, unknown substances can be easily identified and their purity can bé determined. Obviously, this property can be used to distinguish whether two compounds with similar or identical melting points are the same compound. The method is very simple: mix these two compounds together and observe their melting point. If the melting point decreases and the melting range broadens, it indicates that they are two different compounds. It should be noted that a few compounds may undergo decomposition when heated. Therefore, even if their purity is high, they do not have a well-defined melting point and have a wide melting range.

2.6.2 Experimental procedure

Place the dried test sample on a dry and clean surface dish. Then, vertically insert one end of the capillary tube for melting point measurement into the powdered sample, allowing a small amount of the sample to enter the capillary tube. Then, turn the opening end of the capillary tube upwards and gently tap the sealed end of the capillary tube on the experimental bench a few times to allow the sample to settle at the bottom of the capillary tube. Repeat this operation several times, and then turn the sealed end of the capillary tube downwards. Drop the capillary tube freely into a glass tube approximately 50 cm long and upright on the surface dish, and repeat this process several times to ensure that the sample in the capillary tube is compact and uniform; the sample height is about 4 mm. Then, fix the capillary tube containing the sample on the thermometer with a thin rubber band, and place the sample section of the capillary tube at the level of the mercury bulb (see Figure 2-16).

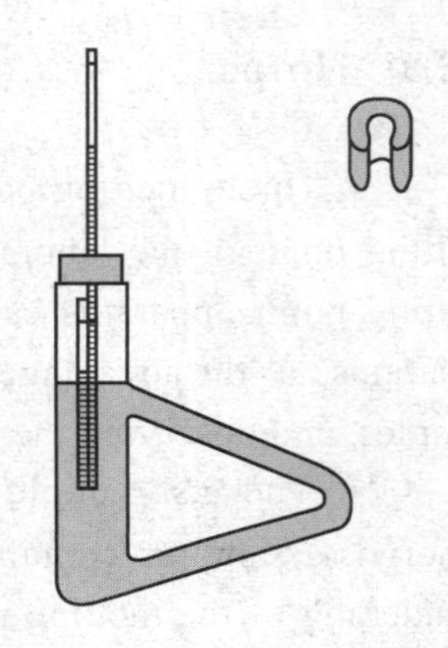

Figure 2-16 Thiele melting point determination tube

Fix the Thiele melting point determination tube on an iron stand, and inject the heat-conducting liquid so that the liquid level of the heat-conducting liquid is at the intersection of the Thiele melting point determination tube. The tube opening is equipped with a soft stopper with a small groove. Insert the thermometer with the melting point capillary tube into it, so that the mercury bulb of the thermometer is located between the two branches of the Thiele melting point determination tube.

During the coarse measurement, heat the bottom of the Thiele melting point determination tube with a small flame, and the heating rate should be around 5°C/min. Carefully observe the temperature changes and whether the sample melts. Record the temperature when the sample melts, which gives the approximate melting point of the sample. Remove the flame and let the temperature of the heat-conducting liquid drop to about 30°C below the approximate melting point for reference in the fine measurement.

During the accurate measurement, remove the thermometer from the Thiele melting point determination tube, replace it with a second melting point tube, and then start heating for measurement. The initial heating can be faster, about 5°C/min; when the temperature reaches about 10°C below the approximate melting point, control the heating rate at around 1°C/min. If the sample in the melting point tube collapses, becomes wet, or even shows small droplets, it indicates the beginning of melting, and the temperature at this time (i.e., initial melting temperature) should be recorded. Continue to slowly heat until the sample is completely melted, and record the temperature at which complete melting occurs (i.e., when the majority of the solid in the tube has melted and only a small amount of fine crystals are about to disappear). The process of solid melting is shown in Figure 2-17.

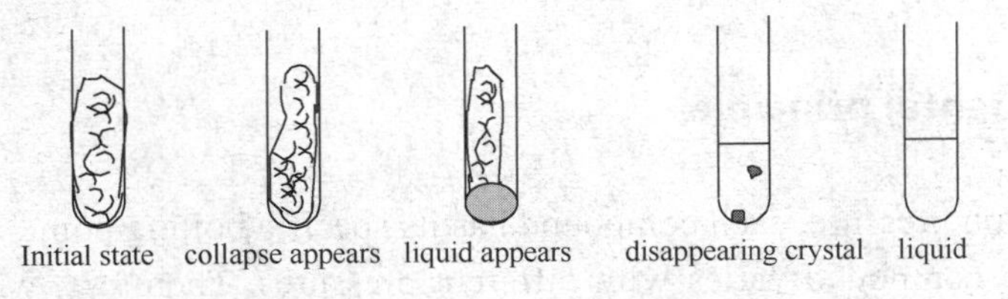

Figure 2-17 Melting process of solid

2.6.3 Notes

(1) Thiele melting point determination tube is a commonly used method for measuring melting point in the laboratory. In addition, microscopic melting point apparatuss or digital melting point apparatus can also be used. Among them, using a microscopic melting point apparatus has the advantages of requiring less sample, being able to measure high melting point samples, and observing the changes of the sample during heating.

(2) The test sample must be thoroughly dried before measuring the melting point. Otherwise, samples containing moisture will cause a decrease in the melting point and a broadening of the melting range. In addition, the sample should be finely ground and packed tightly and uniformly. Otherwise, uneven heat transfer between sample particles can also cause a widening of the melting range.

(3) The choice of heat-conducting medium can be determined based on the melting point of the substance to be tested. If the melting point is below 95℃, water can be used as the heat-conducting liquid. If the melting point is in the range of 95～220℃, liquid paraffin oil can be used. If the melting point is higher, concentrated sulfuric acid (250～270℃) can be used, but caution should be exercised for safety.

(4) When injecting the heat-conducting liquid into the Thiele melting point determination tube, do not use an excessive amount. The expansion of the volume of the heat-conducting liquid after heating should be taken into consideration. In addition, the fine rubber band used to fix the melting point tube should not be immersed in the heat-conducting liquid to prevent dissolution and detachment.

(5) After the test sample has cooled from the melting point measurement, it will revert to a solid state and may form different crystal forms due to different crystallization conditions. Different crystal forms of the same compound often have different melting points. Therefore, a new melting point tube with freshly packed sample should be used for each melting point measurement.

2.7 Boiling Point Determination

When a pure liquid substance is heated until the vapor pressure is equal to the atmospheric pressure, it starts boiling, and the temperature at this point is called the boiling point (bp) of the substance. Boiling point is one of the physical constants of organic compounds, and it can be used to identify and determine the purity of organic compounds.

2.7.1 Experimental principle

Under a certain pressure, each compound has its specific boiling point. In other words, the boiling point of a compound varies with different pressures. Therefore, when describing the boiling point of a compound, the pressure conditions should be specified. For example, the boiling temperature of benzophenone at 13.3 kPa（100 mmHg）is 224.4℃, denoted as 224.4℃/13.3 kPa. However, if referring to the boiling point at normal atmospheric pressure, the pressure conditions are usually not indicated. For example, the boiling point of benzophenone is recorded as 305.4℃, which refers to the boiling point at normal atmospheric pressure. It should be noted that liquids with constant boiling points are not necessarily pure compounds because azeotropic mixtures also have constant boiling points. Therefore, boiling point determination can only qualitatively identify a compound. However, for a known substance, the range of boiling points can be used to determine its purity. Pure compounds generally have a narrow boiling range, usually around 0.5～1℃. Boiling point determination can be carried out using the constant boiling method or the micro-method. The constant boiling method uses a distillation apparatus, which is similar to the process of simple distillation（see 2.2）; while the micro-method uses a similar setup as the melting point determination apparatus.

2.7.2 Experimental procedure

Use a glass tube with an inner diameter of 3～4 mm, a length of 8～10 cm, and one end sealed as the boiling point tube. Add one drop of the liquid to be tested into the tube. Use another glass capillary tube with an inner diameter of approximately 1 mm and a length of about 9 cm as the inner tube, with one end sealed. Insert the open end of the inner tube into the boiling point tube with the bottom of the boiling point tube positioned at the mercury bulb of the thermometer, and insert it into the Thiele melting point determination tube（see Figure 2-16）. Heat slowly and gradually increase the temperature. Soon, bubbles will be observed escaping from the liquid in the boiling point tube due to the expansion of gas in the inner tube caused by heating. When the temperature reaches the boiling point of the liquid, a series of bubbles will rapidly escape from the boiling point tube. Immediately stop heating and let the bath solution cool down naturally. The rate at which gas escapes from the tube will slow down. When the last bubble retracts into the inner tube due to the influx of liquid, the vapor pressure inside the inner tube is equal to the external pressure, and the temperature at this point is the boiling point of the liquid at normal atmospheric pressure.

Figure 2-18 Boiling point determination apparatus

2.7.3 Notes

（1）During boiling point determination, heating should not be too intense, especially when approaching the boiling point of the sample, the temperature should be increased slowly. Otherwise, the liquid in the boiling point tube may evaporate quickly before it can be measured.

（2）If a series of small bubbles escaping rapidly is not observed during the heating process of boiling point determination, it may be due to poor sealing of the inner tube in the boiling point tube. In this case, stop heating, replace the inner tube, and re-determine the boiling point after the temperature of the heat-conducting liquid has decreased by 20℃.

2.8 Recrystallization

The process of separating a substance from impurities based on the difference in solubility between the purified substance and impurities in the same solvent is called recrystallization. Recrystallization is the most commonly used method for purifying solid organic compounds.

2.8.1 Experimental principle

The solubility of solid organic compounds in solvents is greatly affected by temperature. Generally, increasing the temperature increases the solubility, while decreasing the temperature decreases the solubility. If a solid organic compound is dissolved in a hot saturated solution and then cooled, the solubility decreases, resulting in the formation of a cold supersaturated solution, and crystals will precipitate. For the same solvent, different substances have different solubilities. Recrystallization takes advantage of the different solubilities of different substances in the solvent, or removes poorly soluble impurities through hot filtration, or allows well soluble impurities to remain in the mother liquor during the cooling and crystallization process, thereby achieving purification and separation.

2.8.2 Experimental procedure

(1) Constant mass recrystallization

For the purification of solid samples weighing more than 1 g, the constant mass recrystallization method is generally used. First, place the organic compound to be recrystallized in a round-bottom flask, add a smaller amount of solvent, and add a few pieces of boiling stones. Set up a reflux condenser (see Figure 2-4). Connect the condenser to a cold water supply, heat the mixture to boiling, and occasionally shake it. If there are still undissolved solids, gradually add more solvent and maintain reflux. If the boiling point of the solvent is low, after all the solids have dissolved, add some more solvent, which is about 15% of the total amount of solvent added.

If the solution contains colored impurities, activated charcoal can be used for decolorization. Before adding activated charcoal, the solution should be allowed to cool slightly to avoid boiling over. The amount of activated charcoal added is generally 1% to 5% of the weight of the organic compound to be recrystallized. Continue heating and boil for 5 to 10 minutes, then filter the hot mixture through a preheated Büchner funnel to remove insoluble impurities and activated charcoal. Let the resulting filtrate cool naturally to room temperature to allow crystals to precipitate. Then filter at room temperature to remove impurities that have higher solubility in the solvent and still remain in the mother liquor. After filtering off the mother liquor, wash the solid with a small amount of solvent several times, and then dry the crystals on a watch glass. The purity of the crystals can be preliminarily determined by melting point determination.

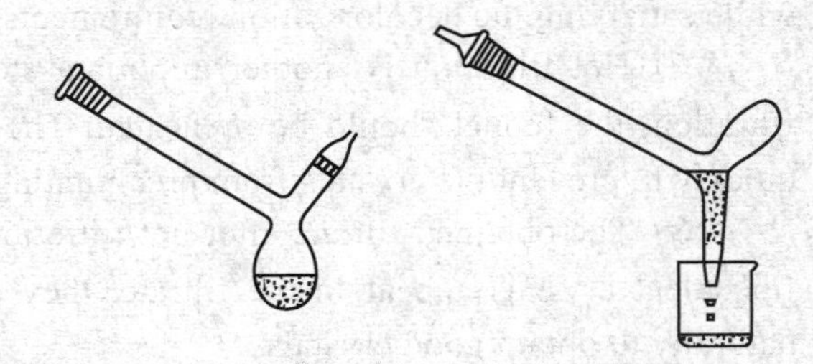

Figure 2-19 Y Sand core funnel filter apparatus

（2）Semi-micro recrystallization: If the amount of the sample to be purified is small（less than 500 mg）, recrystallization using an ordinary Büchner funnel is difficult and usually results in large losses. In this case, using a Y-shaped filter funnel is more convenient and results in smaller product losses（see Figure 2-19）.

First, place the sample in the bulb of the funnel through the glass tube, add a small amount of solvent to rinse the sample inside the glass tube, place the bulb in an oil bath or water bath and heat it to a slight boil, then use a dropper to add solvent to the bulb until the sample is completely dissolved. Stop the heating and wipe off any oil or water stains on the glass bulb. Then, quickly invert the glass bulb and pressurize the hot saturated solution in the funnel through the sand filter into a clean container using a rubber balloon through the glass tube into the Y-shaped funnel. Allow it to settle and crystallize.

2.8.3 Notes

（1）Choosing a suitable solvent is an important step in the recrystallization process. The selected solvent should meet the following conditions: it should not react chemically with the substance to be purified; there should be a significant difference in solubility between the substance to be purified and impurities in the selected solvent, especially the solubility of the substance to be purified should vary significantly with temperature; in addition, the solvent should be easy to separate from the recrystallized material. If the selected solvent satisfies the above conditions and is also economical, safe, low in toxicity, and easy to recover, then it is ideal.

（2）If water is used as the solvent, a reflux apparatus may not be necessary. When using volatile organic solvents, a reflux apparatus is generally required.

（3）When using volatile solvents, excess solvent is usually added to prevent the crystals from precipitating on the filter funnel due to rapid solvent evaporation during hot filtration. Additionally, caution should be exercised when adding flammable solvents to avoid open flames.

（4）The presence of colored impurities in the solution will contaminate the precipitated crystals, and the presence of resinous substances will affect the recrystallization process. In such cases, activated charcoal can be used. Generally, activated charcoal has better decolorization effects in polar solvents（such as aqueous solutions）than in non-polar solvents. It should be noted that while activated charcoal adsorbs impurities, it also has adsorption properties towards the substance being purified. Therefore, the amount of activated charcoal should be minimized while satisfying the decolorization requirements.

(5) Hot filtration is another important step in the recrystallization process. Before hot filtration, the funnel should be preheated. The hot filtration operation should be performed quickly to prevent the crystals from precipitating on the funnel due to temperature drop.

(6) The obtained filtrate after hot filtration should be allowed to cool and crystallize. If flocculent crystals appear in the filtrate, they can be dissolved by heating and then cooled naturally to obtain good crystals.

(7) After cooling, crystallization, and filtration, some crystals may still precipitate from the mother liquor at room temperature, but their purity is lower than the first batch of crystals. If a certain level of crystal purity is required, the first and second batches of crystals should not be mixed together.

(8) Before hot filtration using a Y-shaped tube, the glass ball of the sample solution must be wiped clean to prevent contamination of the filtrate when inverted for filtration.

2.9 Extraction

The process of extracting the desired substance from a solid or liquid mixture using a solvent is called extraction. Extraction is not only a common method for extracting and purifying organic compounds, but it can also be used to remove small impurities from mixtures.

2.9.1 Experimental principle

Extraction is a method that utilizes the property of a substance having different solubilities in two immiscible solvents to transfer it from one solvent to another, achieving the purpose of separation or purification.

At a certain temperature, the same substance (M) follows the following distribution principle in two immiscible solvents (A and B):

$$K = (m_M/V_A) / (m'_M/V'_B)$$

Here, K represents the distribution constant; m_M/V_A represents the amount (g) of component M dissolved in solvent A with a volume of V; m'_M/V'_B represents the amount (g) of component M dissolved in solvent B with a volume of V'.

In other words, the ratio of the solubilities of substance M in two immiscible solvents is a constant at a certain temperature. The above equation can also be written as:

$$K = (m_M/m'_M) \times (V'_B/V_A)$$

It can be seen that when the volumes of the two solvents are equal, the distribution constant K is equal to the ratio of the solubilities of substance M in the two solvents. Obviously, if the volume of the solvent is increased, the amount of substance M dissolved in it will also increase.

From the above formulas, it can also be deduced that multiple sequential extractions are more efficient than a single extraction when using a fixed amount of solvent. However, this does not mean that the more extraction cycles, the higher the efficiency. Generally, extracting three times is recommended, and each time the amount of extractant used is approximately 1/3 of the volume of the solution being extracted.

In addition, the efficiency of extraction is closely related to the choice of solvent. In general, the basic principle for selecting a solvent is that it should have a high solubility for the substance to be extracted, be immiscible with the original solvent, have a low boiling point, and be low in toxicity. For example, when extracting organic compounds from water, solvents such as chloroform, petroleum ether, ethyl ether, and ethyl acetate are commonly used. If it is necessary to remove acids, bases, or other water-soluble impurities from organic compounds, diluted alkali, diluted acid, or water can be used for washing, respectively.

The above discussion is specific to liquid-liquid extraction. If it is necessary to extract certain components from a solid, extraction and separation are carried out based on the different solubilities of the target components and impurities in the same solvent. In the laboratory, a Soxhlet extractor (also known as a fat extractor) is commonly used for continuous extraction from solids. The working principle is to continuously extract solid substances with a solvent by heating and refluxing the solvent and utilizing siphoning.

2.9.2 Experimental procedure

(1) Liquid-liquid extraction

Place the separatory funnel in an iron ring fixed on an iron stand. Pour the mixture to be extracted (volume V) and the extractant (approximately volume $V/3$) into the separatory funnel, and seal it with a stopper. Grip the upper part of the separatory funnel with the right hand and press the stopper with the right index finger; grip the lower part of the separatory funnel with the left hand and oscillate carefully to ensure sufficient contact between the extractant and the mixture to be extracted. During the oscillation, tilt the rear end of the funnel upward from time to time and open the stopcock to release any gas generated by the oscillation (see Figure 2-20). Repeat the oscillation and degassing several times, then place the separatory funnel back on the iron ring and let it stand to allow the layers to separate. After the two phases have clarified, open the stopper on the separatory funnel first, and then open the stopcock to slowly let the lower layer of liquid flow out through the lower end of the funnel, while the upper layer of liquid will flow out from the upper part of the funnel. In this way, the extractant will separate from the original mixture along with the desired substance. Generally, three extractions are sufficient. After merging the extractants, the extract can be obtained by drying and evaporating the solvent through distillation.

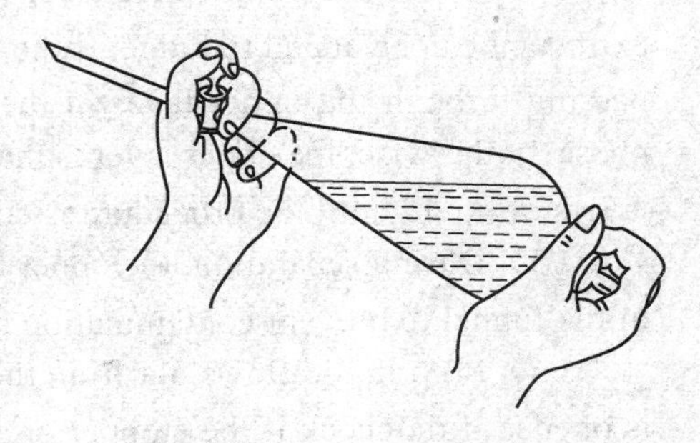

Figure 2-20 Liquid-liquid extraction apparatus

(2) Liquid-solid extraction

Grind the sample to be extracted into a fine powder and wrap it in filter paper, securely fastened with a string, forming a cylindrical shape, and place it in an extraction tube. Add solvent to the round-bottom flask and add a few pieces of boiling stones. Set up a reflux condenser (see Figure 2-21). Start heating to make the solvent boil, and keep the condensate dripping continuously into the extraction tube to accumulate the solvent. When the liquid level is higher than the top of the siphon tube, the extracting solution soaked in the sample will automatically flow back into the flask. The solvent will evaporate when heated and the solvent vapor will reflux into the extraction tube through condensation, repeating this process to continuously accumulate the extract in the flask. Once most of the extract has been obtained, evaporate the solvent to obtain the desired extract.

2.9.3 Notes

(1) The volume of the separatory funnel used should generally be 1～2 times larger than the volume of the liquid being processed. A thin layer of Vaseline should be applied to the piston of the separatory funnel, but not in the piston hole. Then rotate the piston to ensure uniformity and transparency. Before performing the extraction, add an appropriate amount of water to check if there is any leakage in the piston.

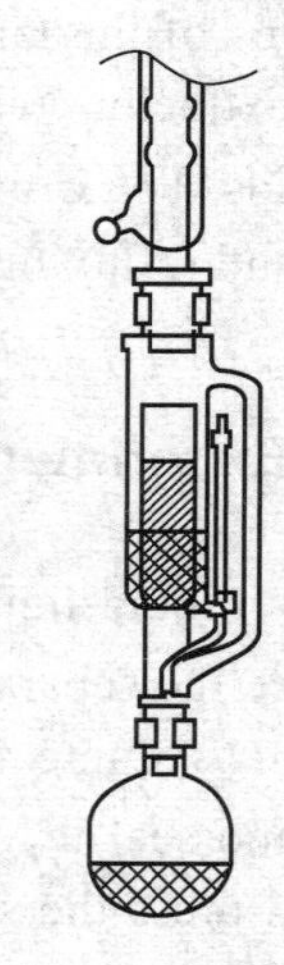

Figure 2-21 Liquid-solid extraction apparatus

(2) When using low-boiling solvents (such as ether) as extractants or washing acid solutions with sodium carbonate solutions, it is important to degas regularly during shaking. Otherwise, the liquid in the separatory funnel may spray out from the stopper.

(3) If emulsification occurs during shaking, demulsification can be achieved by adding a strong electrolyte (such as table salt).

(4) If it is uncertain which layer is the extract phase during separation, a small amount of extractant can be added to judge: if the added extractant dissolves into the lower layer of liquid passing through the upper layer in the separatory funnel, then the lower layer is the extract phase; otherwise, the upper layer is the extract phase. To avoid mistakes, it is best to keep both layers of liquid until the procedure is completed.

(5) During separation, the upper layer of liquid should be poured out from the upper part of the funnel to prevent contamination of the extract layer.

(6) If no liquid flows out from the lower end of the separatory funnel when the stopcock is opened, first check if the stopper on the funnel is open. If the stopper is open and the liquid still cannot flow out, check if the piston hole is blocked.

(7) The most significant advantage of using a Soxhlet extractor for extraction is solvent saving. However, due to the prolonged heating of the extracted substance in the flask, this

method is not suitable for substances that are easily decomposed or discolored by heat. In addition, the boiling point of the solvent used for extraction with a Soxhlet extractor should not be too high.

2.10 Sublimation

Solid substances, when heated, directly transition into vapor without melting. This vapor, upon condensation, directly transforms back into a solid. This process is known as sublimation. Sublimation is a method used for purifying solid organic compounds. It not only separates solid mixtures with different volatilities but also removes non-volatile impurities. Generally, solid organic compounds obtained through sublimation have high purity. However, due to the time-consuming and loss-prone nature of this operation, sublimation is typically limited to the purification of small quantities of substances in the laboratory.

2.10.1 Experimental principle

Broadly speaking, the process of vaporization or evaporation that results in vapor directly transitioning into a solid without passing through a liquid phase is called sublimation. Generally, substances that can be purified through sublimation are solid materials with relatively high vapor pressures below their melting points. These substances have a triple point at which the solid, liquid, and gas phases coexist. The melting point of a substance usually refers to the temperature at which the solid and liquid phases of the substance reach equilibrium under atmospheric pressure. The triple point of a substance refers to the temperature and pressure at which the solid, liquid, and gas phases of the substance reach equilibrium. Below the triple point, the substance exists only in the solid and gas phases. In this case, if the temperature is lowered below the triple point, the vapor can directly transition into the solid state without passing through the liquid state. Conversely, if the temperature is increased, the solid will directly transition into the gas state. Therefore, the process of sublimation should be carried out below the triple point temperature. For example, the triple point temperature of hexachloroethane is 186°C, with a pressure of 104.0 kPa（780 mmHg）. When the temperature is raised to 185°C, the vapor pressure of hexachloroethane reaches 101.3 kPa（760 mmHg）, and it can directly evaporate from the solid phase under normal pressure.

In addition, some substances have relatively low equilibrium vapor pressures at their triple points, resulting in poor sublimation effects under normal pressure. In such cases, sublimation can be performed under reduced pressure conditions.

2.10.2 Experimental procedure

The substance to be sublimed is finely ground and placed in an evaporating dish. Then, a piece of filter paper with many small holes is placed over the mouth of the evaporating dish, and an inverted glass funnel is placed on top of the filter paper. A loose cotton ball is inserted into the neck of the funnel（see Figure 2-22）. The substance in the evaporating dish is slowly sublimed by heating over a low flame through a wire gauze. The vapor rises through the small holes in the filter paper and condenses on the walls of the glass funnel, causing some solid crystals to form on the surface of the filter paper. Once sublimation is complete, the crystals formed on the walls of the funnel and on the filter paper are carefully scraped off and collected using a stainless steel spatula.

The sublimation procedure under reduced pressure is similar to the above procedure for normal-pressure sublimation. First, the substance to be sublimed is placed in a suction filtration tube, and a finger-shaped condenser tube is attached to the filtration tube. Cold water is circulated through the condenser tube, and oil bath heating is used. The opening of the suction filtration tube is connected to a water pump or oil pump（see Figure 2-22）.

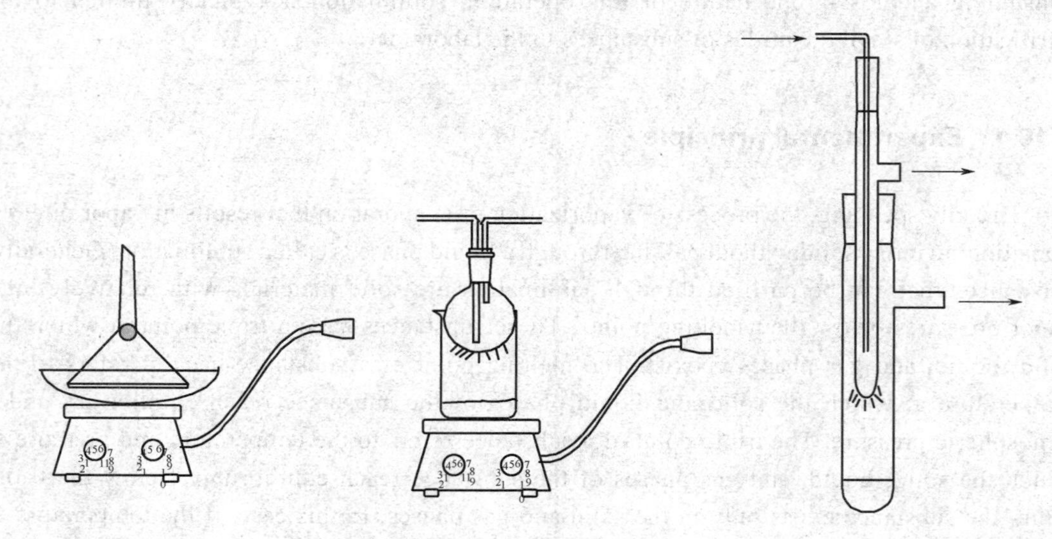

Figure 2-22　Sublimation apparatus

2.10.3　Notes

（1）The substance to be sublimed must be sufficiently dried. Otherwise, during the sublimation process, some organic compounds will evaporate along with water vapor, affecting the separation effect.

（2）Covering the evaporating dish with a layer of filter paper with small holes is mainly to create a temperature difference layer above the evaporating dish, making it easier for the escaping vapor to condense on the walls of the glass funnel and improve the yield of the substance sublimation. If necessary, a cold cloth can be applied to the outer wall of the glass funnel to aid in condensation.

（3）In order to achieve good sublimation separation results, it is better to use a sand bath or oil bath for heating and avoid direct heating with an open flame. The heating temperature should be controlled below the triple point temperature of the substance to be purified. If the heating temperature exceeds the triple point temperature, substances with different volatilities will evaporate together, thereby reducing the separation efficiency.

2.11 Chromatography

Chromatography is one of the important methods for separation, purification, and identification of organic compounds. Chromatography was originally developed for the separation of colored substances, hence its name. Later, with the introduction of various coloration and identification techniques, its application has expanded to include colorless substances.

2.11.1 Experimental principle

There are many types of chromatography, but the basic principles are the same. It utilizes the differences in affinity of various components in the mixture (referred to as the stationary phase) in a certain substance, such as differences in adsorption and solubility (or partitioning), allowing the mixture solution (referred to as the mobile phase) to flow through the stationary phase. This results in repeated adsorption or distribution interactions between the mixture and the mobile phase and stationary phase, thereby separating the various components in the mixture. Depending on different operating conditions, chromatography can be divided into column chromatography, paper chromatography, thin-layer chromatography (TLC). Depending on different mobile phase, chromatography can be divided into gas chromatography and liquid chromatography, etc.

2.11.2 Experimental procedure

(1) Column Chromatography

Select a suitable chromatographic column, wash and dry it, and fix it vertically on an iron stand. Place a filter flask or conical flask at the bottom of the column (refer to Figure 2-23). If there is no sand core separator at the bottom of the column, take a small amount of defatted cotton or glass wool and push it to the bottom of the column using a glass rod, then add a layer of sand approximately 1 cm thick. Close the piston at the bottom of the column, pour solvent into the column to about three-quarters of its height. Then, mix a certain amount of adsorbent (or support agent) with the solvent to form a paste, and add it spoon by spoon from the top of the column into the column. At the same time, open the piston at the bottom of the column

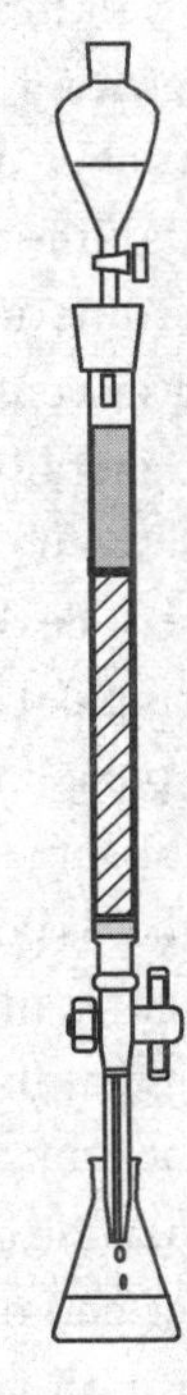

Figure 2-23 Column chromatography

to allow the solvent to flow slowly into the conical flask. During the addition of the adsorbent, gently tap the chromatographic column with a wooden test tube clamp or a glass rod with a rubber tube to promote even settling of the adsorbent. After adding the adsorbent, cover it with a layer of approximately 1 cm thick sand. Throughout the process, the solvent level should always be higher than the adsorbent layer（see Figure 2-23）.

When the solvent level in the column reaches the surface of the adsorbent layer, close the piston at the bottom of the column. Use a dropper to add the prepared sample solution to the surface of the adsorbent layer inside the column. Use the dropper to wash the sample solution adhering to the inner wall of the column with a small amount of solvent. Then, open the piston and allow the solvent to flow out slowly. When the liquid level drops to the level of the adsorbent layer, the eluting solvent can be added for elution. If the separated components have colors, the eluate can be collected based on the appearance of color bands in the chromatographic column. If the components are colorless, they can be collected according to the equal division collection method, and then identified one by one using thin-layer chromatography. The collected fractions of the same component can be combined, and the solvent can be evaporated to obtain the individual components.

（2）Thin-Layer Chromatography

Slowly add 5 g of silica gel G to a 12 mL 1% carboxymethyl cellulose sodium（CMC）solution while stirring to form a paste. Then pour the paste-like slurry onto clean glass plates and gently shake by hand to spread the coating evenly and smoothly. Approximately 6 to 8 glass plates measuring 8 cm × 3 cm can be covered. Air dry at room temperature, then activate in a 110°C oven for 0.5 hours.

Prepare a solution of the sample with a concentration of about 1%. Use a capillary tube with an inner diameter of less than 1 mm to spot the sample. Before spotting, lightly draw a horizontal line 1 cm from the end of the plate with a pencil, then gently spot the sample solution onto the line using the capillary tube. If re-spotting is necessary, make sure to wait for the residual solvent from the previous spotting to evaporate before spotting again to avoid oversized spots. Generally, the spot diameter should not exceed 2 mm. If two samples are spotted on the same plate, the distance between the two spots should be about 1～1.5 cm.

Use a developing tank as the development chamber and add a developing solvent. The solvent level should be about 0.5 cm high. Place a piece of filter paper against the wall of the tank to facilitate the establishment of gas-liquid equilibrium inside the tank. Once the filter paper is completely wetted by the solvent, incline the dried plate and place it in the tank with the spotted end facing downwards, ensuring that the spots are above the level of the developing solvent. Cover the tank with a lid（refer to Figure 2-24）.

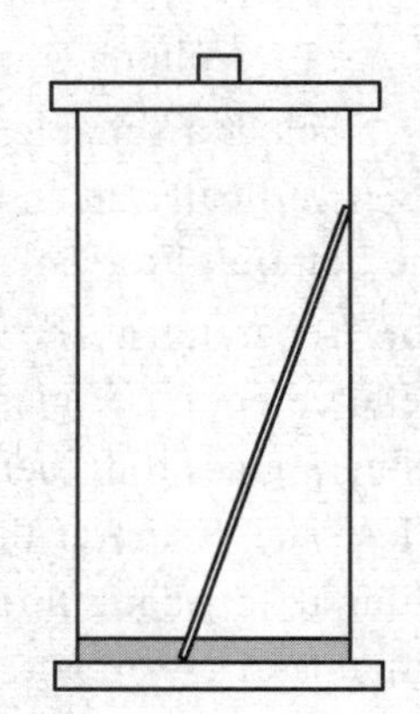
Figure2-24 Thin-layer chromatography

When the developing solvent reaches approximately 1 cm below the upper end of the thin-layer plate, remove the plate and mark the front position of the developing solvent with a pencil. After the thin-layer plate has dried, observe the positions of the spots. If the spots are colorless, place the thin-layer plate in a wide-mouthed bottle containing a few iodine crystals and cover the bottle. When distinct dark brown spots appear on the thin-layer plate, remove it and immediately mark the spot positions with a pencil. Then, calculate the R_f values of each spot.

(3) Gas Chromatography

Select a clean and dry stainless steel tube (or sometimes a glass tube) with a suitable length as the chromatographic column. Take a support material slightly more than the volume of the column, and take an amount of fixing liquid equal to 5% to 25% of the mass of the support material. Mix the fixing liquid with a low-boiling point solvent (such as ether, acetone, or chloroform) in an amount equivalent to the support material and stir evenly. Then, remove the solvent using a rotary evaporator (or heat with an infrared lamp) to leave behind the fixed liquid on the support material. Place the support material impregnated with the fixing liquid into a 110～120℃ oven to age for 2 hours. Plug one end of the selected chromatographic column with a glass wool and connect it to a vacuum pump, while connecting a small funnel to the other end. Start the vacuum pump and gradually pour the aged support material into the funnel, allowing the support material to be sucked into the column. During the column packing process, continuously tap and vibrate the chromatographic column to ensure uniform and compact filling of the support material. Once finished, remove the funnel and plug the end of the chromatographic column with a glass wool. Use this ends as the inlet and connects it to the chromatograph (see Figure 2-25).

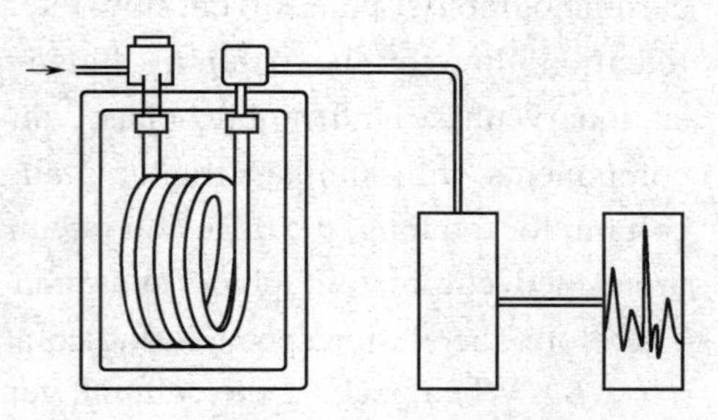

Figure 2-25 Gas chromatography

Install the chromatographic column in the column compartment of the chromatograph, then power on the instrument and adjust the carrier gas flow rate (about 10～5 mL/min) and operating temperature (slightly higher than the desired temperature but lower than the maximum temperature for the fixing liquid). Once the baseline of the recorder becomes stable, the sample can be injected for analysis.

2.11.3 Notes

(1) When separating a mixture by chromatography, factors such as the properties of the adsorbent, polarity of the solvent, size of the column, amount of adsorbent used, and elution speed should be considered.

(2) The selection of adsorbents generally depends on the type of compounds to be separated. For example, acidic alumina is suitable for separating acidic compounds such as carboxylic acids or amino acids; basic alumina is suitable for separating amines; neutral alumina can be used for separating neutral compounds. Silica gel has mild properties, is amorphous and porous, slightly acidic, and suitable for separating substances with high polarity, such as alcohols, carboxylic acids, esters, ketones, and amines.

(3) The choice of solvent depends on the polarity and solubility of the compounds to be

separated. Sometimes, a single solvent can separate all components in a mixture; other times, a mixture of solvents or alternating solvents may be required. For example, a non-polar solvent can be used first to elute non-polar components from the column, followed by a polar solvent to elute polar components. Common solvents (in increasing polarity) include petroleum ether, carbon tetrachloride, toluene, dichloromethane, chloroform, acetic acid, ethyl acetate, acetone, ethanol, methanol, water, etc.

(4) The size of the chromatographic column and the amount of adsorbent used depend on the quantity and separation difficulty of the sample to be separated. Generally, the ratio of column length to column diameter is about 8∶1, and the amount of adsorbent is about 30 times the mass of the sample to be separated. After packing the adsorbent into the column, approximately one-fourth of the column capacity should be left for solvent. Of course, if the separation is more challenging, a longer column and more adsorbent can be used.

(5) The flow rate of the solvent significantly affects the separation efficiency of column chromatography. If the solvent flow rate is slow, the components in the mixture will have longer retention time in the column, allowing sufficient adsorption or distribution between the stationary phase and mobile phase, thus enabling the separation of the mixture, especially components with similar structures and properties. However, if the components stay in the column for too long, the diffusion rate in the solvent may exceed the rate of elution, resulting in broadened chromatographic bands and overlapping peaks that affect separation efficiency. Therefore, the elution speed during column chromatography should be moderate.

(6) When packing the column, gently tap the column continuously to remove air bubbles and avoid leaving cracks, as they can affect separation efficiency.

(7) After packing the column, when adding solvent to the column, slowly pour it along the column wall to avoid splashing and covering the surface of the adsorbent and samples with sand particles.

(8) In addition to separation and purification, thin-layer chromatography can also be used for the identification of organic compounds and finding suitable column chromatography conditions. In organic synthesis, it can be used to monitor reaction progress. The principle of separation is based on the capillary action of the adsorbent on a thin layer of plate in the mobile phase, causing the components of the sample mixture to ascend with the mobile phase. Due to differences in adsorption on the adsorbent and solubility in the mobile phase, separation occurs during ascension. Under certain chromatographic conditions, the ratio of the distance traveled by a compound to the distance traveled by the mobile phase is a constant value called the relative front value (R_f value). It is an important criterion for comparing and identifying different compounds. However, it should be noted that the reproducibility of R_f values is poor in practical work. Therefore, in the identification process, known substances and unknown substances are often spotted on the same thin-layer plate and developed in the same mobile phase. By comparing their Rf values, judgment can be made.

(9) Common adsorbents used in thin-layer chromatography are silica gel and alumina. Silica gel without binders is referred to as silica gel H; silica gel with binders such as calcined gypsum is called silica gel G; silica gel with fluorescent substances is called silica gel HF254, which can be observed under ultraviolet light at a wavelength of 254 nm, while organic

compounds attached to the bright fluorescent thin-layer plate appear as dark spots, allowing observation of colorless components. Silica gel GF254 refers to silica gel containing both calcined gypsum and fluorescent substances. Alumina can also be divided into alumina G, alumina HF254, and alumina GF254. Due to the strong polarity of alumina, it has a strong adsorption effect on polar substances, making it suitable for separating weakly polar compounds such as hydrocarbons, ethers, and halogenated hydrocarbons. Silica gel has relatively lower polarity, making it suitable for separating highly polar compounds such as carboxylic acids, alcohols, and amines.

(10) During plate preparation, adsorbents should be gradually added to the solvent while stirring. Adding the solvent to the adsorbent in reverse order may cause clumping.

(11) When spotting samples, ensure that the opening of the capillary tube is flat, and perform the spotting action quickly and agilely. Otherwise, oversize spots, trailing, diffusion, and other phenomena may occur, affecting separation efficiency.

(12) The polarity difference of the mobile phase significantly affects the separation of mixtures. If all components of a mixture with strong polarity move forward with the solvent during chromatography, it means that the polarity of the solvent is too strong. On the contrary, if the spots of all components in the mixture do not move along with the elution of the solvent, it means that the polarity of the solvent is too weak. When choosing a mobile phase, refer to point (3). It should be noted that sometimes a single solvent is not sufficient for separating a mixture, which requires a mixture of solvents as the mobile phase. The polarity of this mixed mobile phase is often between the polarities of several pure solvents. A rapid way to find a suitable mobile phase is as follows: spot several spots of the sample to be separated on a thin-layer plate with a spacing of more than 1 cm between them. Use a dropper to apply different solvents to different spots. These spots will expand into concentric circles with different sizes as the solvent spreads. By observing the interlayer distance between these circles, the suitability of the solvent can be roughly determined.

(13) Iodine staining is an effective method for observing colorless spots. Iodine can form colored complexes with most organic compounds except alkanes and halogenated hydrocarbons. However, because iodine sublimes, the stained spots on the thin-layer plate will disappear after the plate is exposed to the air for some time. Therefore, after iodine staining, the stained spots should be immediately circled with a pencil. If the thin-layer plate contains fluorescent substances, the compounds can be directly observed under ultraviolet light, and the compounds will appear as dark spots due to the absorption of ultraviolet light.

(14) Gas chromatography is a chromatographic method that uses gas as the mobile phase (carrier gas). According to the state of the stationary phase, it can be divided into gas-solid chromatography and gas-liquid chromatography. The experimental method described here is gas-liquid chromatography. Gas-liquid chromatography uses a porous inert solid material as the support (also known as the substrate), which is coated with a thin layer of high-boiling organic liquid compound as the stationary phase (also known as the liquid stationary phase) and filled in the chromatographic column. When the carrier gas carries the mixture into the chromatographic column, the components of the mixture will repeatedly distributed between the carrier gas and the liquid stationary phase. Those components with low solubility in the liquid stationary phase will be quickly carried out by the carrier gas, while components with

high solubility in the liquid stationary phase will move slowly, resulting in the separation of the components. Gas-solid chromatography is similar to gas-liquid chromatography in principle. The difference is that gas-solid chromatography directly uses some porous solid adsorbents such as silica gel and activated alumina as the stationary phase.

(15) There are many models of gas chromatographs, but their components are basically the same. They mainly include the carrier gas supply system, sample introduction system, chromatographic column, detection system, and recording system, etc. The operating conditions depend on the specific model used. Generally, once the chromatograph is stable, a microsyringe can be used for sample injection. The vaporized sample will be separated into individual components in the chromatographic column and sequentially enter the detector, where the detector will convert these components at different concentrations into electrical signals and record them as chromatographic peaks on a recorder. Usually, the time required from the start of sample injection to the appearance of the maximum concentration of a component is called the retention time. Generally, under the same analytical conditions, the retention time of an organic compound remains constant. Therefore, gas chromatography can be used for qualitative analysis. In addition, the content of each component is proportional to its peak area, so quantitative analysis can also be performed based on peak areas.

(16) Hydrogen gas is used in the operation of gas chromatography. Open flames should be avoided to ensure safety.

2.12 Determination of Refractive Index

The refractive index is one of the physical constants of organic liquid compounds. By measuring the refractive index, the purity of organic compounds can be determined, and it can also be used to identify unknown substances.

2.12.1 Experimental principle

In different media, the speed of light propagation is not the same. When light enters another medium from one medium, its direction of propagation will change. This is the phenomenon of refraction. According to the law of refraction, when a light ray passes from medium A to medium B, the ratio of the sine of the incident angle α to the sine of the refracted angle β is inversely proportional to the refractive indices of the two media:

$$\sin\alpha / \sin\beta = n_B / n_A$$

If medium A is set as the optically sparse medium and medium B as the optically dense medium, then $n_A < n_B$. In other words, the refracted angle β must be smaller than the incident angle α, as shown in Figure 2-26.

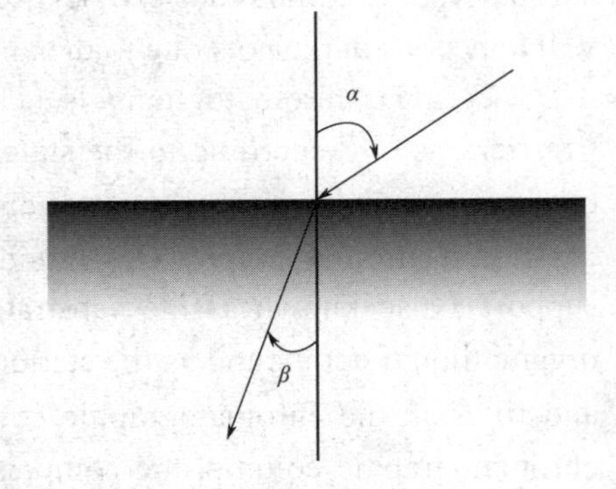

Figure 2-26 Phenomenon of refraction

If the incident angle α is 90°, i.e., $\sin\alpha = 1$, the refracted angle is at its maximum value (referred to as the critical angle, denoted as β_0). The determination of the refractive index is usually carried out in air, but it can still be approximately considered in a vacuum state, i.e., $n_A = 1$.

$$n=1/\sin\beta_0$$

Therefore, by measuring the critical angle β_0, the refractive index n of the medium can be obtained. Typically, the refractive index is measured using an Abbe refractometer, which is based on the phenomenon of light refraction.

Since factors such as the wavelength of the incident light and the measurement temperature significantly affect the refractive index of a substance, the measured value is usually annotated with the operating conditions. For example, at a temperature of 20°C, the refractive index of carbon tetrachloride measured using sodium light D-line wavelength (589.3 nm) is 1.4600, denoted as n_D^{20} 1.4600. Since the measured data can be read to the fourth decimal place, it has high precision and good repeatability, making the refractive index even more reliable as a purity standard for liquid organic substances than boiling point. Additionally, the influence of temperature on the refractive index is inversely proportional. Typically, for every 1°C increase in temperature, the refractive index decreases by 3.5×10^{-4} to 5.5×10^{-4}. For convenience, the temperature change constant is often approximated to 4×10^{-4} in practical work. For example, the measured value of methyl tert-butyl ether at 25°C is 1.3670, and its corrected value should be: $n_D^{20}=1.3670+5\times4\times10^{-4}=1.3690$.

2.12.2 Experimental procedure

Open the prism of the refractometer (refer to Figure 2-27), first wipe the mirror surface of the prism with lens paper soaked in acetone, then add 1-2 drops of the test sample onto the prism surface, and close the prism. Rotate the reflection mirror to allow the light to enter the prism, making both eyepieces bright. Then adjust the prism adjustment knob until a semi-bright and semi-dark pattern can be observed in the eyepiece. If a color band appears, adjust the dispersion prism (micro-adjustment knob) to make the boundary line clear. Next, adjust the brightness boundary line to precisely coincide with the center of the crosshair in the eyepiece (refer to Figure 2-28). Record the readings and temperature, repeat the process twice, and take the average value. After the measurement is completed, open the prism and wipe the mirror surface clean with acetone.

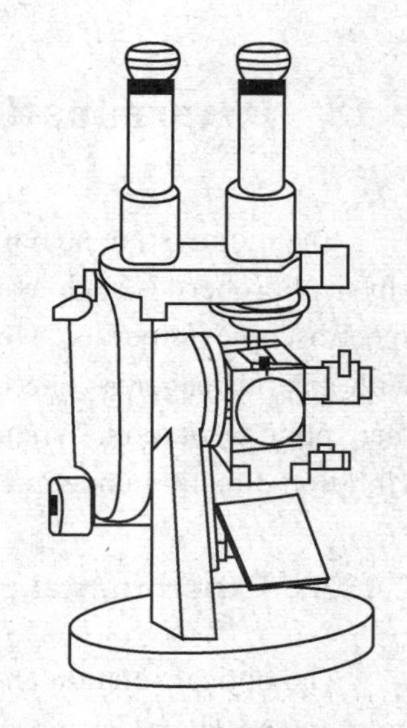

Figure 2-27 Refractometer

2.12.3 Notes

(1) Since the Abbe refractometer is equipped with a dispersion prism, it can convert polychromatic light into monochromatic light. Therefore, it is possible to directly use daylight to measure the

refractive index, and the obtained data is the same as that obtained using sodium light.

(2) Pay attention to protecting the prism of the refractometer and avoid measuring corrosive liquids such as strong acids or strong alkalis.

(3) Before the measurement, the prism must be wiped with lens paper dipped in a volatile solvent to remove any residue that may affect the measurement results.

(4) If measuring volatile liquids, the sample can be added through a small hole on the side of the prism.

(5) Common situations when measuring refractive index are shown in Figure 2-28, where Figure 2-28 (4) is the pattern for reading data. When encountering Figure 2-28 (1), which indicates the appearance of a dispersion color band, the micro-adjustment knob of the prism must be adjusted until the color band disappears and presents the pattern in Figure 2-28 (2), then adjust the prism adjustment knob until it presents the pattern in Figure 2-28 (4). If encountering Figure 2-28 (3), it is due to insufficient sample quantity, so more samples should be added and measured again.

(6) If the field of view inside the reading eyepiece is unclear, check whether the small reflection mirror is open.

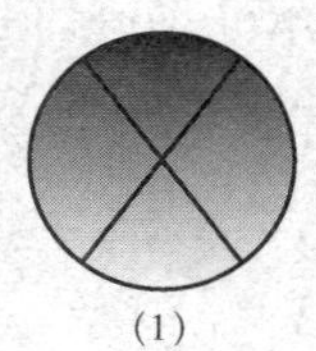
(1)

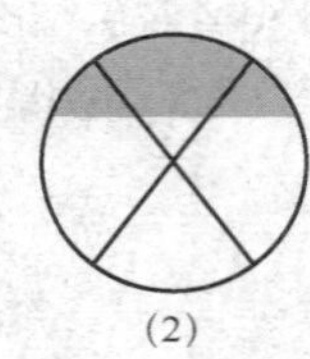
(2)

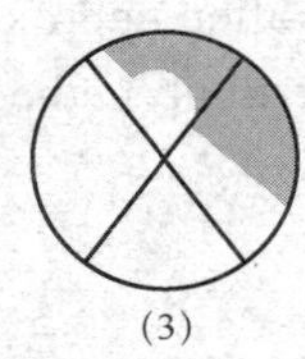
(3)

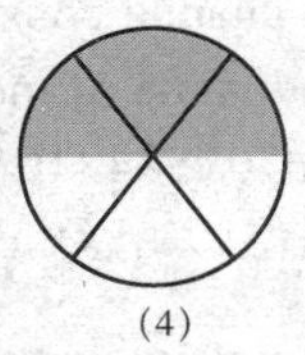
(4)

Figure 2-28 Pattern inside the refractometer

2.13 Determination of Optical Rotation

Enantiomers are stereoisomers that are mirror images of each other. They have the same physical properties such as melting point, boiling point, relative density, refractive index, and spectroscopic properties. Moreover, their chemical properties are also identical when interacting with achiral reagents. The only property that reflects the difference in molecular structure is their optical rotation. When polarized light passes through an optically active substance, its vibration direction undergoes rotation, and the angle of rotation is called optical rotation.

2.13.1 Experimental principle

The optical rotation and direction of optically active substances can be measured using a polarimeter. The polarimeter mainly consists of a sodium light source, two Nicol prisms, and a sample cell containing the test sample (see Figure 2-29). Ordinary light passes through a fixed prism (polarizer) to become polarized light, then passes through the sample cell, and finally its vibration direction and rotation angle are examined by a rotatable prism (analyzer). If the polarized light's vibration plane rotates to the right, it is called dextrorotatory; if it rotates to the left, it is called levorotatory.

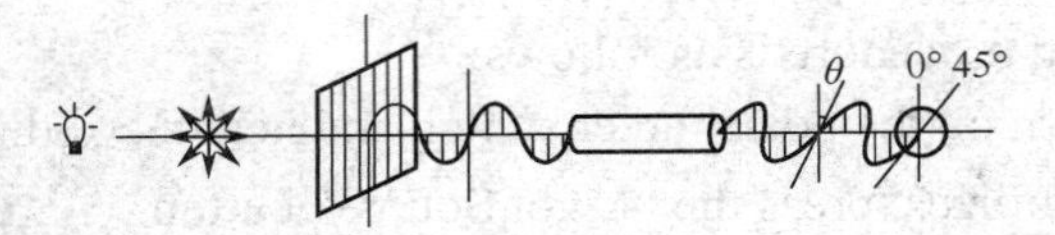

Figure 2-29 Polarized light

The optical rotation of optically active substances is closely related to factors such as concentration, test temperature, and wavelength of light. However, under certain conditions, the optical rotation of each optically active substance is a constant, represented by specific rotation angle（$[\alpha]$）:

$$[\alpha]^t_\lambda = \alpha / (c \cdot l)$$

Here, α represents the measurement value from the polarimeter; c represents the concentration of the sample solution, expressed as the weight of the sample in grams per milliliter of solution; l represents the length of the sample cell in decimeters; λ represents the wavelength of the light source, usually represented by D-line sodium light. t represents the test temperature. If the sample being tested is in liquid form, it can be directly measured without the need to prepare a solution. When calculating the specific rotation angle, the density value（d）can be substituted for the concentration value（c）in the formula:

$$[\alpha]_\lambda^t = \alpha / (d \cdot l)$$

In addition to specific rotation angle, other parameters such as optical purity, percentage content of levorotatory and dextrorotatory enantiomers, and enantiomeric excess（e.e）can be used to reflect the purity of optically active substances. If S represents the main enantiomer content in a mixture of enantiomers, R represents the other enantiomer content, the enantiomeric excess（e.e）is calculated using the following formula:

$$\text{e.e.\%} = [(s - R) / (s + R)] \times 100$$

If the optical purity of the (−)-enantiomer is x%, then

$$\text{Percentage content of (−)-enantiomer} = [x + (100 - x) / 2] \times 100\%$$

$$\text{Percentage content of (+)-enantiomer} = (100 - x) / 2 \times 100\%$$

Optical purity（P）is defined as:

$$P = ([\alpha]^t_D \text{(sample)} / [\alpha]^t_D \text{(standard)}]) \times 100\%$$

For example, if the sample of（S）-(−)-2-methylbutanol has a relative density of d_4^{23} =0.8 and exhibits an optical rotation of -8.1 in a sample cell with a length of 20 cm, and the standard sample has $[\alpha] = -5.8$（pure）, then:

$$\text{Specific rotation angle } [\alpha]^t_D \text{(sample)} = \alpha^{23} / (c \cdot l) = -8.1 / (2 \times 0.8) = -5.1$$

$$\text{Optical purity } P = [[\alpha]^t_D \text{(sample)} / [\alpha]^t_D \text{(standard)}] \times 100\% = (-5.1) / (-5.8) \times 100\% =$$

$$88\% \text{ Percentage content of (−)-enantiomer} = [88 + (100 - 88) / 2] \times 100\% = 94\%$$

$$\text{Percentage content of (+)-enantiomer} = (100 - 88) / 2 \times 100\% = 6\%$$

$$\text{Enantiomeric excess e.e.\%} = (S - R) / (S + R) \times 100\% = 88\%$$

2.13.2 Experimental procedure

There are various types of polarimeters. Taking the example of a digital automatic display polarimeter, the operating procedure is as follows:

(1) Preheating: Turn on the polarimeter switch and heat the sodium lamp for 15 minutes. After the light source stabilizes, press the "Light Source" button.

(2) Zero adjustment: Fill the sample cell with the solvent used to prepare the test sample solution or distilled water, and place the sample cell in the testing chamber to adjust to zero, making the reading on the digital display (or scale) zero.

(3) Preparation of solution: Accurately weigh 0.1-0.5 g of the sample and dissolve it in a 25 mL volumetric flask. Water, ethanol, or chloroform can be used as solvents. If testing a pure liquid sample, determine its relative density before testing.

(4) Testing: Use a suitable length of sample cell, fill it with the sample solution or pure liquid sample, and remove air bubbles. Then, place the sample cell in the testing chamber and close the cover. Press the "Measure" button and wait for the reading on the digital display (or scale) to stabilize. Take two more measurements and calculate the average value. Calculate specific rotation angle, enantiomeric excess, etc., using the formulas.

2.13.3 Notes

(1) If the specific rotation angle value of the sample is small, it is advisable to prepare a higher concentration of the test sample solution and use a longer sample cell for better observation.

(2) Temperature changes have a certain influence on optical rotation. If testing under sodium light (λ=589.3 nm), the optical rotation of most optically active substances decreases by approximately 0.3% per 1°C temperature increase.

(3) During testing, the position of the sample cell should remain fixed to eliminate testing errors caused by changes in distance.

第三部分

有机化合物合成通法

3.1 卤化反应

3.1.1 卤化反应原理

向有机分子中引入卤素原子以制备卤代烃，这个过程叫做卤化反应。卤代烃不仅是重要的有机合成中间体，也是常用的有机溶剂。从卤代烃可以衍生出许多有应用价值的化合物，如醇、酚、醚、胺、醛、酮、酸等。因此，卤化反应在有机合成中应用十分广泛。

卤代烃分子中烃基结构的不同，制备的方法和所用的卤化剂也有差异。如脂肪烃卤代物可用卤化氢与醇反应而制得，也可以用卤素分子与脂肪烃在光照条件下进行自由基取代反应而获得。芳烃卤代物可以用氯、溴作卤化剂，在铁粉或相应的三卤化铁催化下与芳烃发生亲电取代而制得。

$$C_6H_5NO_2 + Br_2 \xrightarrow{Fe} m\text{-}BrC_6H_4NO_2$$

由于氟与芳烃的直接反应过于激烈甚至会导致芳烃分解，因此，氟代芳烃通常是由相应的芳胺经重氮化、桑德迈耶（Sandmeyer）反应来制备。如果直接用碘与芳烃作用，虽然可获得碘代芳烃，但由于同时伴生的碘化氢对碘代芳烃具有还原作用，使反应具有可逆性。如：

$$ArH + I_2 \rightleftharpoons ArI + HI$$

显然，若向反应体系中添加氧化剂如浓硫酸或硝酸，就能使碘代反应正向移动。若以卤代烃作为合成中间体，在卤代烃的实验室制备过程中，以选择溴化最为适宜，因为溴在室温下为液体，易于操作。

3.1.2 芳烃溴化实验通法

对于 0.1 mol 芳烃投料量，可用 250 mL 三口烧瓶作反应容器，配置搅拌器、回流冷凝管、滴液漏斗、温度计以及溴化氢气体吸收装置（见图 2-6、图 2-7）。

芳烃的溴化条件与芳烃的反应活性密切相关。

如果芳烃反应活性较低，先将 0.1 mol 芳烃和 0.5 g 铁粉加到三口烧瓶中，在油浴中加热至 100～150℃。然后，在搅拌下经滴液漏斗向三口烧瓶滴加 10 g（0.06 mol）溴，加溴速度以不使溴蒸气通过冷凝管逸出为宜。滴加完毕，继续搅拌并加热反应 1 h，再加入 0.5 g 铁粉，滴加 10 g 溴。加料完毕，继续搅拌并加热反应 2 h。

将含有少量溴的红棕色反应混合物倒入事先溶有 1 g 亚硫酸氢钠的 200 mL 水溶液中，以除去残余的溴。若溴色还未除尽，可再加少许亚硫酸氢钠。通常在溴化产物中总会残留一些未反应的溴，用水不易洗净，若用亚硫酸氢钠则可使其还原成水溶性的溴化钠而去除：

$$Br_2 + 3NaHSO_3 \longrightarrow 2NaBr + NaHSO_4 + 2SO_2 + H_2O$$

然后，对混合物进行水蒸气蒸馏。产物若为固体，过滤晾干后即可满足一般使用的要求，必要时可进行重结晶。产物若为液体，可用四氯化碳提取；若产物为不溶于水的液体，可直接用分液漏斗分离。然后，分别用 10%氢氧化钠水溶液、水洗涤至中性，用硫酸镁干燥后蒸除溶剂，对残余物做减压蒸馏。

当芳烃反应活性居中，先将 0.1 mol 芳烃和 0.2 g 铁粉加到三口烧瓶中，然后在室温下边搅拌边滴加 16 g（0.1 mol）溴。加溴完毕，在室温下搅拌反应 1 h，反应过程中有溴化氢气体放出。如果反应过于缓慢，可用水浴（30～40℃）加热反应 1 h。然后，将水浴温度提高到 65℃左右，继续搅拌反应一段时间，直到反应混合物液面不再有红棕色蒸气逸出为止。后处理同前。

当芳烃反应活性较高，溴化就必须在温和的条件下进行。先用 50 mL 四氯化碳将 0.1 mol 芳烃稀释并冷却至 0℃，然后边搅拌边滴加 0.08 mol 溴的四氯化碳溶液（由 10 mL 四氯化碳与 13 g 溴配制而成），反应温度控制在 0～5℃。加溴完毕，仍保持在低温下（0～5℃）搅拌 1～2 h，直到反应混合物液面无红棕色蒸气逸出为止。后处理同前。

实验中必须注意溴具有强腐蚀性，对皮肤有很强的灼伤性，其蒸气对黏膜有刺激作用，因此在量取时必须带上橡皮手套并在通风橱中进行。

量取溴时，先将放置在铁圈上的分液漏斗安放在通风橱内，然后把溴倒入分液漏斗，再用量筒经分液漏斗量取溴。

3.1.3 实验

实验一　对–溴乙酰苯胺

【实验目的】

学习芳烃卤化反应理论，掌握芳烃溴化方法，熟悉溴的物理、化学特性及其使用操作方法。掌握重结晶及熔点测定技术。

【实验原理】

NHCOCH$_3$ (苯环) $\xrightarrow{Br_2/HAc}$ NHCOCH$_3$ (对位 Br 取代苯环)

【药品】

乙酰苯胺	13.5 g（0.1 mol）
溴	16 g（5 mL，0.1 mol）
冰醋酸	36 mL
亚硫酸氢钠	1～2 g

【实验操作】

在 250 mL 三口烧瓶上，配置搅拌器、温度计、滴液漏斗和回流冷凝管，回流冷凝管连接气体吸收装置以吸收反应中产生的溴化氢。

注意：搅拌器与三口烧瓶口连接处的密封性要好，以防溴化氢从瓶口溢出。

向三口烧瓶中加入 13.5 g 乙酰苯胺和 30 mL 冰醋酸，用温水浴稍稍加热，使乙酰苯胺溶解。然后，在 45℃温水浴条件下，边搅拌边滴加 16 g 溴和 6 mL 冰醋酸配成的溶液。滴加速度以棕红溴色能较快褪去为宜。

注意：溴具有强腐蚀性和刺激性，必须在通风橱中量取。操作时应戴上橡皮手套。

滴加完毕，在 45℃浴温下继续搅拌反应 1 h，然后将浴温提高至 60℃，再搅拌一段时间，直到反应混合物液面不再有红棕色蒸气逸出为止。

将反应混合物倾入盛有 200 mL 冷水的烧杯中（如果产物带有棕红色，可事先将 1 g 亚硫酸氢钠溶入冷水中；如果产物颜色仍然较深，可适量再加一些亚硫酸氢钠）。用玻璃棒搅拌 10 min，待反应混合物冷却至室温后过滤，用冷水洗涤滤饼并抽干，在 50～60℃温度下干燥，产物可以直接用于对-溴苯胺的制备。

对-溴乙酰苯胺可以用甲醇或乙醇重结晶。产物经干燥后，称重、测熔点并计算产率。对-溴乙酰苯胺为无色晶体，mp 为 164～166℃。

【注意事项】

（1）室温低于 16℃时，冰醋酸呈固体，可将盛有冰醋酸的试剂瓶置入温水浴中融化。

（2）滴速不宜过快，否则反应太剧烈，会导致一部分溴来不及参与反应就与溴化氢一起逸出，同时也可能会产生二溴代产物。

【思考题】

（1）乙酰苯胺的一溴代产物为什么以对位异构体为主？

（2）在溴化反应中，反应温度的高低对反应结果有何影响？

（3）在对反应混合物的后处理过程中，用亚硫酸氢钠水溶液洗涤的目的是什么？

（4）产物中可能存在哪些杂质，如何除去？

3.2　磺化反应

3.2.1　磺化反应原理

有机分子中的氢原子被磺酸基（$—SO_3H$）所取代的反应称为磺化反应。芳烃的磺化反应属亲电取代反应，是一类应用极广的单元反应。在芳环上引入磺酸基可以增强水溶性，这在染料、医药合成方面有着重要意义。由于磺酸基可以方便地转化为羟基、氨基、硝基、氰基等，因而磺酸类化合物又是有机合成中的重要中间体。另外，许多芳烃的磺化产物自身就具有重要的应用价值，例如，具有 12～15 个碳原子烷基的烷基苯磺酸盐可用作洗涤剂。较低级的烷基萘磺酸盐是应用广泛的润湿剂、乳化剂，如二丁基萘磺酸钠。

常用的磺化剂有浓硫酸、氯磺酸（$ClSO_3H$）、三氧化硫等。其中，硫酸是最温和的磺化剂，通常用于磺化较活泼的芳烃。氯磺酸属较剧烈的磺化剂，它不仅可以磺化芳烃，还可磺化脂肪烃。芳烃与等物质的量的氯磺酸作用，生成的是芳磺酸，若与过量的氯磺酸反应，生成的是芳磺酰氯。三氧化硫是最强的磺化剂，如发烟硫酸（三氧化硫溶于硫酸中）可用来磺化低活性的芳烃，不过用三氧化硫作磺化剂容易发生氧化反应，因而宜在较低温度下进行磺化反应，或以卤代烃作稀释剂使反应缓和。

磺化产物磺酸与硫酸类似，属水溶性强酸，能溶于过量的磺化剂中。磺化反应结束后，通常先用冰水将反应混合物稀释，

NH_2 + H_2SO_4 $\xrightarrow{185℃}$ NH_2 / SO_3H

NO_2 + $ClSO_3H$ $\xrightarrow{105℃}$ NO_2 / SO_2Cl

再用碱中和并加入饱和食盐水使磺酸以盐的形式析出（盐析法）：

$$ArSO_3H + NaCl \rightleftharpoons ArSO_3Na\downarrow + HCl$$

与硫酸不同的是，磺酸的钙盐、钡盐都溶于水，利用这一差别也可用碳酸钙（钡）中和，滤去硫酸盐沉淀，以除去过量的硫酸，滤液再用碳酸钠溶液处理，滤去生成的碳酸钙盐沉淀，就可得到磺酸钠盐（磺酸钠盐的这种纯化法也称脱硫酸钙法）。

$$\left.\begin{matrix} H_2SO_4 \\ ArSO_3H \end{matrix}\right\} + CaCO_3 \longrightarrow (ArSO_3)_2Ca + CaSO_4$$

以氯磺酸作磺化剂，其磺化产物磺酰氯微溶于水，分离纯化要简便一些，很多磺酰氯都可以通过蒸馏加以提纯，磺酰氯经水解就可得到磺酸。此外，由于磺酰氯十分活泼，由它可以制备出许多有用的磺酸衍生物，如磺酰胺、磺酸酯等。因此在实验室中，氯磺酸用得更普遍一些。

3.2.2 芳烃氯磺化实验通法

对于 0.1 mol 的芳烃投料量，可用 250 mL 三口烧瓶作反应容器，配置搅拌器、回流冷凝管、滴液漏斗、温度计以及气体吸收装置（见图 2-6 和图 2-7）。

芳烃氯磺化的反应条件与芳烃反应活性有密切的关系。

如果芳烃反应活性较低，可将 0.3 mol 氯磺酸和 0.1 mol 芳烃一同加入三口烧瓶，加热搅拌，慢慢升温至 110～120℃，此时液面有大量氯化氢逸出。当氯化氢气体逸出趋缓，反应已近结束，可将反应温度提高 10℃，继续搅拌，直至无氯化氢气体放出。

若芳烃反应活性高，为了避免反应过于激烈，可用 25 mL 干燥氯仿将 0.1 mol 芳烃先行稀释，用冰盐浴将其冷却至−10℃左右，在激烈搅拌下，慢慢滴加 0.2 mol 氯磺酸。此时，有大量的氯化氢气体放出。滴毕，仍在−10℃左右继续搅拌。当氯化氢气体逸出趋缓，可将

反应混合物温热至室温，并继续搅拌直到不再放出氯化氢气体。

若芳烃反应活性居中，可先加 0.1 mol 芳烃于烧瓶中并冷却至 0℃，在激烈搅拌下，滴加 0.25 mol 氯磺酸。滴毕，在室温下继续搅拌，直至不再有氯化氢气体逸出。

反应结束后，在搅拌下于通风橱中将反应混合物慢慢倒入 100 g 碎冰中，析出的磺酰氯若为固体产物，可以进行过滤、洗涤和重结晶操作；若为液体产物，可以用氯仿或苯等溶剂对反应混合物萃取，然后依次用水、碳酸氢钠水溶液和水洗涤，最后进行蒸馏，即可获得氯磺化产物。

注意：氯磺酸和浓硫酸类似，具有强酸性和强腐蚀性，会烧伤皮肤，操作时应戴上橡皮手套，在通风橱中操作。

3.2.3　芳烃磺化实验通法

对于 0.1 mol 芳烃投料量，可用 100 mL 三口烧瓶作反应容器，配置搅拌器、回流冷凝管和温度计（见图 2-6）。

如果芳烃反应活性较低，可将三口烧瓶置于冰水浴中，加入 35 g 25%发烟硫酸，在搅拌下滴加 0.1 mol 芳烃，此时混合物发热，注意冷却控温，防止过热。芳烃滴加完毕，在一定温度下继续搅拌 1～2 h。

若芳烃反应活性高或居中，都可以 30 g（0.3 mol）浓硫酸作磺化剂，其磺化过程与低活性芳烃的磺化过程相同。磺化完毕，静置冷却，在搅拌下将反应混合物沿烧杯壁慢慢倒入 100 mL 冰水中。待混合液冷却后，小心加入碳酸钠，使溶液呈中性，然后加入 30 g 氯化钠，使芳烃磺酸盐逐渐析出。过滤析出的沉淀，用少量水洗涤、抽干，置于烘箱中于 50～130℃干燥。

【注意事项】

（1）磺化温度要取决于具体的磺化对象以及磺酸基进入芳环上的位置，温度差异比较大。例如，在 80℃左右，萘磺化反应的主要产物为 α-萘磺酸；在 180℃左右反应，主要产物是 β-萘磺酸。

（2）具体干燥温度要视水合芳烃磺酸盐脱水温度而定。

（3）浓硫酸具有强吸水位和强腐蚀性，对皮肤有灼伤性，操作应当心。

3.2.4　实验

实验二　邻-甲苯磺酰氯和对-甲苯磺酰氯

【实验目的】

学习氯磺化反应的原理及实验方法，熟悉使用气体吸收装置及重结晶和减压蒸馏操作。

【实验原理】

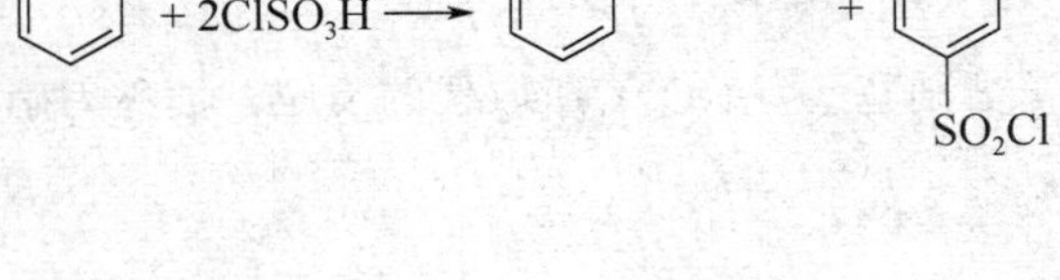

$$C_6H_5CH_3 + 2ClSO_3H \longrightarrow o\text{-}CH_3C_6H_4SO_2Cl + p\text{-}CH_3C_6H_4SO_2Cl$$

邻-甲苯磺酰氯为无色油状液体，bp126℃/1.3 kPa（10 mmHg）。

对-甲苯磺酰氯为片状晶体，mp67～69℃，bp135℃/1.3 kPa（10 mmHg）。

【药品】

氯磺酸	38 g（22 mL，0.33 mol）
甲苯	10 g（12 mL，0.11 mol）

【实验操作】

在 100 mL 三口烧瓶上配置搅拌器、温度计、滴液漏斗、回流冷凝管及氯化氢气体吸收装置（见图 2-6 及图 2-7）。向三口烧瓶中加入 22 mL 氯磺酸，并置反应瓶于冰浴中冷却至 0℃。

注意：氯磺酸具有强腐蚀性，遇水会猛烈放热甚至爆炸，在空气中就会冒出大量氯化氢气体。因此，反应装置和药品要充分干燥，操作时要当心，应在通风橱中量取。

在搅拌下，自滴液漏斗向反应瓶中滴加 12 mL 无水甲苯。滴速以保持反应温度不超出 5℃为宜。温度过高，会使生成的甲苯磺酰氯发生水解。控制滴速，充分搅拌，大约 15 min 滴毕，继续在室温下搅拌 1 h。然后在 40～50℃温水浴中加热搅拌，直至不再有氯化氢气体放出为止。

待反应液冷却至室温后，在通风橱内，边搅拌边将反应液慢慢倒入盛有 80 mL 冰水的烧杯中，再用 20 mL 冰水洗涤反应瓶，洗涤液并入烧杯。然后，用倾泻法倾出酸层，将淡黄色油状液体分离出来，即得邻-和对-甲苯磺酰氯的混合物。用冰水对混合物洗涤两次后，将油状混合物置入−10～20℃冰柜中冷却（也可以用氯化钙冰盐浴冷却）过夜。

冷冻后，对-甲苯磺酰氯结晶从混合物中析出，抽滤（最好用砂芯漏斗），用少量冷水洗涤滤饼，再抽滤即得对-甲苯酰氯粗品。滤液中主要含邻-甲苯磺酰氯，用氯仿将其萃取，萃取液经水洗后，分出有机相，并用无水硫酸镁干燥。蒸除溶剂后，进行减压蒸馏，收集 126℃/1.3 kPa（10 mmHg）馏分。也可以用石油醚（30～60℃）对产物进行重结晶。

【注意事项】

（1）甲苯沸点为 110.8℃，与水形成共沸物，在 84.1℃沸腾，含 81.4%甲苯。甲苯可以采用共沸蒸馏法进行干燥，把最初 20%的蒸馏液弃去即可。若含水量小，也可以通过加入无水氯化钙来干燥。

（2）将反应物置于冰箱中过夜，有可能提高邻-甲苯磺酰氯的产量。

（3）有时由于对-甲苯磺酰氯在混合物中所占比例不高，其结晶不易析出，此时可以作邻-甲苯磺酰氯粗品处理。

（4）磺酰氯在进行减压蒸馏前，一定要作充分干燥，否则在高温条件下，磺酰氯会发生水解。

【思考题】

（1）如果在氯磺化反应前，所加药品未作干燥，将对反应产生什么影响？

（2）在氯磺化反应结束后，为什么要将反应混合物倒入冰水中？

（3）本实验基于什么原理来分离邻-甲苯磺酰氯、对-甲苯磺酰氯混合物？

（4）如果以甲苯、氯磺酸为原料合成甲基苯磺酸，你将如何对本实验步骤进行修改？

3.3 硝化反应

有机分子中的氢原子被硝基（—NO_2）所取代的反应，被称为硝化反应。芳环上的硝化反应是一类重要的亲电取代反应，在精细有机合成中有着广泛的应用。

在硝化反应中，硝酸阳离子 NO_2^+ 是亲电试剂。质子的存在有助于 NO_2^+ 的生成。

显然，除去水分子有利于提高 NO_2^+ 的浓度，增强硝化活性。事实上，加入浓硫酸既能提供上述平衡式中所需要的质子，又可吸收水分，从而使 NO_2^+ 浓度迅速上升。

$$H^+ + HNO_3 \rightleftharpoons [H_2NO_3]^+$$

$$[H_2NO_3]^+ \rightleftharpoons H_2O + NO_2^+$$

$$HNO_3 + 2H_2SO_4 \rightleftharpoons NO_2^+ + H_2O + 2HSO_4^-$$

不过，对于一些容易硝化的芳环，只需要用稀硝酸作为硝化剂即可，而且硝化条件也比较温和，这类芳环常含有第一类（活化）取代基（如—OH、—OCH_3、—$NHCOCH_3$、—CH_3 等）；带有第二类（钝化）取代基（如—NO_2、—COOH、—SO_3H 等）的芳环，硝化条件要剧烈一些，必须使用由发烟硝酸、浓硫酸配成的混酸作为硝化剂。

NO_2 / NO_2 + 发烟硝酸（硫酸） $\xrightarrow[5天]{110℃}$ NO_2 / O_2N / NO_2

H_3C / CH_3 / H_3C / CH_3 + 发烟硝酸(硫酸) $\xrightarrow[15min]{50℃}$ NO_2 / H_3C / CH_3 / H_3C / CH_3 / NO_2

3.3.1　芳烃硝化实验通法

对于不同反应活性的芳烃，所需要的硝化剂组成是不一样的。以 0.1 mol 芳烃化合物为例，对于高反应活性芳烃，如酚类、苯基醚等，只需 20 mL 40%硝酸（d=1.25，0.2 mol）即可；中等反应活性芳烃，如苯、甲苯、萘等，需 8 mL 68%浓硝酸（d=1.41，0.13 mol）和 10 mL 浓硫酸组成的混酸；低反应活性芳烃，即含有钝化基团的芳烃，如苯甲酸、硝基苯等，需 10 mL 100%浓硝酸（d=1.51，0.24 mol）和 14 mL 100%浓硫酸（d=1.83，0.26 mol）组成的混酸。

注意：硝酸和硫酸有强腐蚀性，使用时要小心谨慎，量取硝酸时，要在通风橱中操作。多硝基化合物绝不可以蒸馏，即使是蒸馏一硝基化合物，也要当心，不能蒸干，以免发生爆炸。

混酸的配制：在冷水浴条件下，将浓硫酸慢慢加入到浓硝酸中，边加边搅拌，并将混酸冷却至 10℃。

在 250 mL 三口烧瓶上，配置搅拌器、温度计和滴液漏斗等，注意反应系统与大气相通（参见图 2-6）。将 0.1 mol 待硝化芳烃加入到三口烧瓶中，搅拌下自滴液漏斗滴入混酸，反应温度一般控制在 20～50℃，反应时间约为 2～3 h。对于高反应活性的芳烃，反应温度应低一些（0～10℃），当混酸滴加完毕后，在室温下搅拌 30 min 即可。

反应完毕，在搅拌下将反应物慢慢倒入 200 mL 冰水中。在室温下如果产物为固体，经过滤、水洗至中性，干燥后即得硝化粗产物，再选择适当溶剂对粗产物重结晶；如果产物为液体，则先用分液漏斗分出有机层，酸液层用乙醚萃取，萃取液与有机相合并，依序用水、

10%碳酸氢钠水溶液、水洗涤至中性，经氯化钙干燥后作减压蒸馏。

3.3.2 实验

实验三 邻–硝基苯酚和对–硝基苯酚

【实验目的】

学习芳烃硝化反应的基本理论和硝化方法，加深对芳烃亲电取代反应的理解，掌握水蒸气蒸馏技术。

【实验原理】

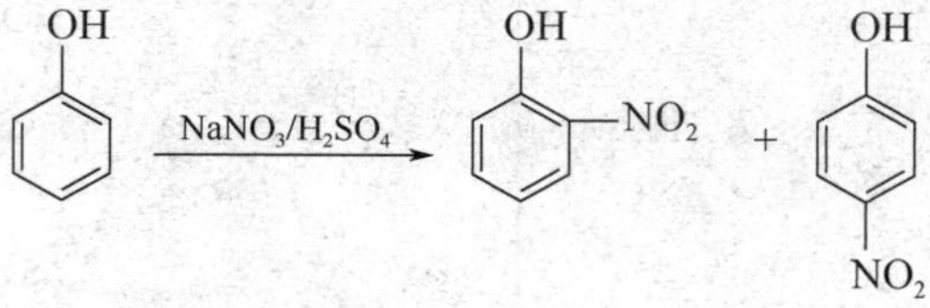

$$C_6H_5OH \xrightarrow{NaNO_3/H_2SO_4} o\text{-}NO_2C_6H_4OH + p\text{-}NO_2C_6H_4OH$$

邻-硝基苯酚 mp45℃，有特殊的芳香气味。

对硝基苯酚为淡黄或无色针状晶体，无气味，mp112～113℃。

【药品】

苯酚	4.7 g（0.05 mol）
硝酸钠	7 g（0.08 mol）
浓硫酸（d=1.83）	11 g（6 mL，0.11 mol）
浓盐酸	3 mL

【实验操作】

在 100 mL 三口烧瓶上，配置搅拌器、温度计和滴液漏斗（参见图 2-6）。先加入 20 mL 水，然后，在搅拌下慢慢加入 6 mL 浓硫酸。

注意：只可将浓硫酸沿容器壁往水中慢慢倾倒，切不可颠倒次序!

取下滴液漏斗，趁酸液尚在温热之时，自反应瓶侧口加入 7 g 硝酸钠，使其溶入稀硫酸中。装上滴液漏斗，将反应瓶置入冰水浴中，使混合物冷却至 20℃。

称取 4.7 g 苯酚，与 1 mL 温水混合，并冷却至室温。

注意：苯酚有腐蚀性，若不慎触及皮肤，应立刻用肥皂和水冲洗，再用酒精棉擦洗。

在搅拌下，将苯酚水溶液自滴液漏斗滴入反应瓶中，用冰水浴将反应温度维持在 20℃左右。

注意：在非均相反应体系中，保持良好的搅拌能够显著地加速反应。

加完苯酚后，在室温下继续搅拌 1 h，有黑色油状物生成，倾出酸层。然后向油状物中加入 20 mL 水并振摇，先倾出洗液，再用水洗三次，以除净残存的酸。

注意：硝基酚产物有毒，洗涤操作时要小心！

对油状混合物作水蒸气蒸馏，直到冷凝管中无黄色油滴馏出为止。在水蒸气蒸馏过程中，黄色的邻-硝基苯酚晶体会附着在冷凝管内壁上，可以通过间或关闭冷却水龙头，使热蒸汽将其熔化而流出。

将馏出液冷却过滤，收集浅黄色晶体，即得邻-硝基苯酚产物。晾干后称量、测熔点并计算产率。

注意：邻-硝基苯酚容易挥发，应保存在密闭的棕色瓶中。

向水蒸气蒸馏后的残余物中加水至总体积为 50 mL，并加入 3 mL 浓盐酸和 0.5 g 活性炭，煮沸 15 min，用预热过的布氏漏斗过滤，滤液经冷却析出对-硝基苯酚。过滤干燥后称重、测熔点并计算产率。

如果实测熔点偏低，可以用乙醇-水混合溶剂对产物进行重结晶：加少量乙醇于盛有一硝基苯酚的圆底烧瓶中，配置回流冷凝管，加热回流，再补加乙醇直到产物全部溶解于沸腾的乙醇中。然后，逐滴加入热水（60℃左右），直到乙醇溶液中正好出现混浊为止。再加几滴乙醇，使混浊液刚好澄清。静置冷却至室温，过滤即得产物，晾干后测熔点。

【注意事项】

（1）苯酚的熔点为 41℃，室温下呈固态，量取时可用温水浴使其熔化。苯酚中加入少许水可降低熔点，使其在室温下即呈液态，有利于滴加和反应。

（2）反应温度对苯酚的硝化影响很大。当温度过高，一元硝基酚有可能发生进一步硝化，或因发生氧化反应而降低一元硝基酚的产量；当温度偏低，又将减缓反应速度。

（3）硝基酚在残余混酸中进行水蒸气蒸馏时，会因长时间高温受热而发生进一步硝化或氧化。因此，一定要洗净粗产物中的残酸。

【思考题】

（1）苯酚的硝化可能会有哪些副反应？

（2）为什么邻-硝基苯酚、对-硝基苯酚可以采用水蒸气蒸馏来分离？可否用同样方法来分离邻-硝基甲苯、对-硝基甲苯？

3.4　傅-克反应实验通法

3.4.1　傅-克反应原理

1877 年，巴黎大学化学家傅列德尔（C. Friedel）和美国化学家克拉夫茨（J. M. Crafts）在合作研究中发现，在氯甲烷和苯的稳定混合溶液中，一旦加入无水三氯化铝，就会发生剧烈反应，并释放出大量氯化氢气体。他们从反应混合物中分离得到了甲苯。后来的研究表明，这类反应具有普遍意义。它不仅适用于芳烃的烷基化反应，而且可用于芳烃的酰基化反应。这就是后来被人们所命名的 Friedel-Crafts 烷基化和酰基化反应，简称为傅-克反应。

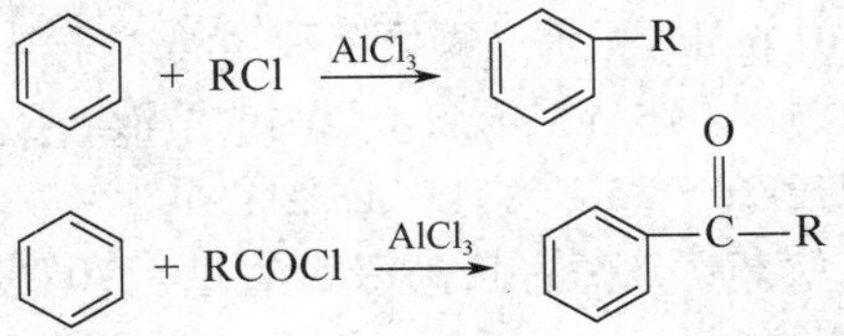

$$C_6H_6 + RCl \xrightarrow{AlCl_3} C_6H_5{-}R$$

$$C_6H_6 + RCOCl \xrightarrow{AlCl_3} C_6H_5{-}\overset{O}{\overset{\|}{C}}{-}R$$

芳烃的烷基化反应和酰基化反应被用来合成烷基芳烃和芳酮，应用十分广泛。

在烷基化反应中，除了卤代烃外，烯烃以及醇类也可作为烷基化试剂。

由于烷基对芳烃具有活化作用，在傅-克烷基化反应中，生成物烷基芳烃比原料芳烃更容易发生烷基反应，从而产生多烷基芳烃。因此，欲制取单烷基芳烃，必须加入过量很多的芳烃以控制多烷基芳烃的形成。在这里，芳烃既作反应试剂，又作稀释剂。如果芳烃是固体，可以另外加入溶剂，如二硫化碳、石油醚或硝基苯等惰性介质。此外，由于芳烃的烷基化反应是通过烷基正碳离子的形成与进攻而发生的，因而会产生重排产物。这就导致烷基化反应的应用在合成多于 2 个碳以上的直链烷基芳烃时，受到一定的限制。

与烷基化反应不同，由于酰基对芳环具有钝化作用，因而芳烃的酰基化反应会停留在一取代阶段。这对于选择性地制备单取代基芳烃是十分有利的；又由于在酰基化反应中不会发生重排，因而在合成直链烷基芳烃或带其他支链结构的烷基芳烃时具有特殊应用价值。例如，正丙苯就可以经酰基化反应和还原反应而制得，它不仅可以停留在一取代阶段，而且能保持原有碳链的构造。

$$C_6H_6 + CH_3CH_2COCl \xrightarrow{AlCl_3} C_6H_5COCH_2CH_3 \xrightarrow{[H]} C_6H_5CH_2CH_2CH_3$$

常用的酰基化试剂有酰氯、酸酐。

在以卤代烃作烷基化试剂的傅-克反应中，常用的催化剂有无水三氯化铝、氯化锌、三氟化硼等路易斯酸，其中尤以无水三氯化铝催化效能最好（在以烯烃或醇类烷基化反应中，一般用质子酸作催化剂，如氟化氢、硫酸以及磷酸等）。在烷基化反应中，无水三氯化铝的投入量仅需催化剂量，但在酰基化反应中，情况就不同了。由于无水三氯化铝可以与羰基化合物形成稳定的配合物，因而仅用催化剂量的无水三氯化铝是不够的。以酰氯作酰基化试剂，考虑到酰氯及产物芳酮都会与三氯化铝形成配合物，因此，1 mol 酰氯投入量，需配以多于 1 mol 无水三氯化铝的投入量，一般过量 10%。若以酸酐作酰基化试剂，则需要更多的无水三氯化铝。因为，酸酐在傅-克反应中，会生成乙酸，乙酸和酰基化产物芳酮一样，都要消耗等物质的量的三氯化铝，以形成配合物。因此，1 mol 酸酐至少需要 2 mol 的三氯化铝，在实际制备中，通常还要过量 10%。在酰基化反应中，时常以过量的芳烃或二硫化碳、硝基苯和石油醚等作溶剂。

3.4.2 烷基化实验通法

就苯的单烷基化反应而论，反应时苯是大大过量的，故以卤代烃投入量作参考对象。

对于 0.1 mol 卤代烃投料量，可用 100 mL 三口烧瓶作为反应容器，配置机械搅拌器、滴液漏斗、温度计和回流冷凝管，冷凝管上端附设氯化钙干燥管，干燥管与气体吸收装置相连（参见图 2-6、图 2-7）。

依次将研碎的 1.3 g（0.01 mol）无水三氯化铝和 30 mL 干燥苯加入三口烧瓶中，搅拌下自滴液漏斗滴入 0.1 mol 卤代烃与 15 mL 干燥苯配制的溶液。反应时自行放热，注意控制滴加速度，滴速以保持反应温度 20～25℃为宜，必要时可采用冰水浴冷却。加料完毕，在室温下继续搅拌直至反应趋于缓和。然后，提高浴温至 60～70℃，加热并搅拌，直到反应混合液中不再有氯化氢气体逸出为止。

待反应混合物冷却后，在通风橱内将其慢慢倒入 50 mL 混有碎冰的冰水中，如果仍有沉淀物，可滴加少量盐酸使其溶解，并用玻璃棒充分搅拌。然后用分液漏斗分除水相，再依次用水、5%碳酸钠水溶液和水对有机相进行洗涤，直至有机相呈中性。经无水氯化钙干燥后，蒸除溶剂，然后通过分馏（或重结晶）收集产品。

3.4.3 酰基化实验通法

对于 0.1 mol 芳烃投料量，可用 250 mL 三口烧瓶作为反应容器，配置机械搅拌器、滴液漏斗和回流冷凝管，冷凝管上端附设氯化钙干燥管，干燥管与气体吸收装置相连（参见图 2-6、图 2-7）。

依次将研碎的 16 g（0.12 mol）无水三氯化铝、100 mL 干燥过的石油醚（60～90℃）和 0.1 mol 芳烃加入到三口烧瓶中。

在搅拌下，慢慢滴加 0.1 mol 酰氯，滴速以控制反应温度在 40℃左右为宜。滴加完毕，在 50～60℃的水浴中，加热并搅拌 1～2 h，直到反应混合液中不再有氯化氢气体逸出为止。

待反应混合物冷却后，在通风橱内慢慢将其倾入 100 mL 冰水中，有氢氧化铝沉淀析出。在搅拌下加入 5～10 mL 浓盐酸，使沉淀物溶解。然后用分液漏斗分出有机相；以石油醚或二氯乙烷对水相提取两次。合并有机相，再依次用水、5%氢氧化钠水溶液和水将有机相洗涤至中性。

经无水硫酸镁干燥后，蒸除溶剂，减压蒸馏收集产品。

若以酸酐作酰基化试剂，其酰化实验步骤与酰氯酰基化类似，只是无水三氯化铝的投入量要增加，即以 0.1 mol 酸酐替代酰氯，应加入 32 g（0.24 mol）无水三氯化铝。

3.4.4 实验一

实验四 二苯甲酮（酰基化法）（富有甜味的香料）

【实验目的】

学习傅-克酰基化反应理论及实验方法，掌握萃取、蒸馏、减压蒸馏等操作技术。

【实验原理】

$$C_6H_5-\overset{O}{\overset{\|}{C}}-Cl + C_6H_6 \xrightarrow{AlCl_3} C_6H_5-\overset{O}{\overset{\|}{C}}-C_6H_5$$

二苯甲酮呈无色晶体，mp47～48℃，bp305.4℃。

【药品】

无水三氯化铝	7.5 g（0.056 mol）
无水苯	27 g（30 mL，0.34 mol）
苯甲酰氯	7.3 g（6 mL，0.05 mol）
5%氢氧化钠水溶液	20 mL
浓盐酸	2～3 mL

【实验操作】

在 100 mL 三口烧瓶上，配置搅拌器、恒压滴液漏斗、温度计和回流冷凝管，冷凝管上端依次配置氯化钙干燥管和盛有碱液的气体吸收装置（参见图 2-6、图 2-7）。

注意：实验中所用仪器和药品均需干燥。

在通风橱内称取 7.5 g 无水三氯化铝，在研钵内研细后迅速投入三口烧瓶中，加入 30 mL 苯。在室温下边搅拌边自滴液漏斗向三口烧瓶内滴加 6 mL 苯甲酰氯。滴速以控制反应温度处于 40℃为宜。

注意：苯甲酰氯具有催泪刺激性，对皮肤、眼睛及呼吸道都有强刺激作用。因此，应在通风橱中量取。

瓶内混合物开始激烈反应，并伴有氯化氢气体产生，反应液逐渐呈褐色。大约 10 min 后滴加完毕，在 60℃水浴上加热并搅拌，直至反应混合物液面不再有氯化氢气体逸出为止，需时 1.5 h 左右。

待三口烧瓶冷却后，在通风橱内将反应物慢慢倒入盛有 50 mL 冰水的烧杯中，有沉淀物析出。搅拌下，用滴管慢慢加入 2～3 mL 浓盐酸，直至沉淀物全部分解。用分液漏斗分出有机相，以苯作萃取剂对水相提取两次（2×15 mL）。合并有机相，依次用 20 mL 水和 20 mL 5%氢氧化钠水溶液对有机相进行洗涤，然后再用水洗涤 2～3 次（每次 20 mL），直至有机相呈中性。经无水硫酸镁干燥、蒸除溶剂，即得粗产物。然后减压蒸馏，收集 187～190℃/2.0 kPa（15 mmHg）馏分，冷却后固化，即得纯品。

如果不经减压蒸馏，粗产物可用石油醚（60～90℃）重结晶。干燥后称重、测熔点并计算产率。

【注意事项】

（1）重新蒸馏苯、弃去 10%初馏分，即可满足要求。

（2）无水三氯化铝极易吸潮，与潮湿空气接触会产生刺激性的氯化氢气体，因此称量、研磨及投料等操作要迅速。

（3）粗产品常呈黏稠状，这是由于溶剂未除尽或有不同晶型的存在，导致熔点下降。

（4）二苯甲酮有多种晶型，它们的熔点各不相同：α 型为 49℃；β 型为 26℃；γ 型为 45～48℃；δ 型为 51℃。其中 α 型晶型较稳定。

【思考题】

（1）用酰氯作酰基化试剂进行傅-克反应时，为什么要用过量许多的无水三氯化铝作催化剂？

（2）酰基化反应结束后为什么要用酸处理？

（3）在酰基化反应中，是否容易产生多酰基取代芳烃？

3.4.5 实验二

实验五　二苯甲酮（烷基化法）

【实验目的】

学习傅-克烷基化反应的理论和实验方法，掌握水蒸气蒸馏、萃取、减压蒸馏等操作技术。

【实验原理】

$$C_6H_6 \xrightarrow[AlCl_3]{CCl_4} C_6H_5-CCl_2-C_6H_5 \xrightarrow{H_2O} C_6H_5-CO-C_6H_5$$

【药品】

无水苯	7.8 g（9 mL，0.1 mol）
四氯化碳	34 g（22 mL，0.22 mol）
无水三氯化铝	6.7 g（0.05 mol）
无水硫酸镁	1～2 g
苯	20 mL

【实验操作】

在 250 mL 三口烧瓶上，配置机械搅拌器、滴液漏斗、温度计和回流冷凝管，冷凝管上端依次附设氯化钙干燥管和气体吸收装置（参见图 2-6、图 2-7）。

注意：所用仪器和药品均需事先干燥。

依次将研细的 6.7 g 无水三氯化铝和 15 mL 四氯化碳迅速投入三口烧瓶中。

注意：四氯化碳有毒！避免吸入其蒸气。无水三氯化铝极易吸潮，操作要快捷。

将三口烧瓶置于冰水浴中，待瓶内温度降至 12℃左右，在搅拌下先慢慢滴加 4 mL 由 9 mL 无水苯和 7 mL 四氯化碳配成的溶液。反应开始后，有氯化氢气体产生，反应混合物温度逐渐升高，此时可用冰水浴将反应温度控制在 12℃。

当反应变得较温和后，将余下的苯溶液逐滴加入到反应烧瓶中，滴速以保持反应温度在 5～10℃之间为宜。滴加完毕（需时约 15 min），继续搅拌 1 h，反应温度保持在 10℃左右。反应完毕，将反应装置改为水蒸气蒸馏装置（见图 2-11）。通过滴液漏斗将 40 mL 水慢慢滴入三口烧瓶中，反应混合物逐渐变热。注意控制水的滴加速度，以保持过量的四氯化碳平稳沸腾并使其直接蒸馏出来。加完水后，让三口烧瓶在石棉网上小火加热 0.5 h，蒸除残余的四氯化碳（如果烧瓶中的水蒸发过多，可由滴液漏斗向瓶中适量加水），并促使二氯二苯甲烷水解完全。然后，将三口烧瓶中的混合物转入分液漏斗，分出有机相，水相用 20 mL 苯萃取。萃取液与有机相合并，合并液用水洗涤至中性，经硫酸镁干燥，常压蒸馏蒸除溶剂，再作减压蒸馏，收集 187～190℃/2.0 kPa（15 mmHg）馏分。产物冷却后为固体，mp47～48℃。

【注意事项】

（1）四氯化碳和苯通过简单蒸馏，弃去 10%的初馏分，就可获得满足傅-克反应要求的无水四氯化碳和无水苯。

（2）当反应温度低于 5℃时，反应太慢；温度高于 10℃时，则易产生焦油状树脂产物。

（3）由于中间体二氯二苯甲烷的水解需要受热，故在分解三氯化铝配合物时，不必冷却，使中间体得以初步水解。

【思考题】

（1）傅-克反应中，烷基化反应和酰基化反应在催化剂无水三氯化铝的使用上各有何不同？为什么？

（2）本实验可能有哪些副反应，为了减少副反应，实验中采取了什么措施？

（3）从理论上讲，本实验每制备 1 mol 二苯甲酮，需要 2 mol 苯和 1 mol 四氯化碳，实际投料比又如何？试简要说明。

3.5 氧化反应

3.5.1 氧化反应原理

化学反应中，凡失去电子的反应称为氧化反应。有机化合物的氧化反应表现为分子中氢原子的减少或氧原子的增加。在有机合成中，氧化反应是一类重要的单元反应，通过氧化反应，可以制取许多含氧化合物，例如醇、醛、酮、酸、酚、醌以及环氧化物等，应用十分广泛。

工业上常以廉价的空气或纯氧作氧化剂，但由于其氧化能力较弱，一般要在高温、高压的条件下才能发生氧化反应；实验室中常用的氧化剂有高锰酸钾、重铬酸钠、硝酸等。这些氧化剂氧化能力强，可以氧化多种基团，属于通用型氧化剂。

以高锰酸钾作氧化剂，在不同的介质中，反应结果不一样。高锰酸钾在中性或碱性介质中进行氧化时，锰原子的价态由+7 下降为+4，生成二氧化锰，它不溶于水而沉淀下来，在这个过程中，平均 1 mol 高锰酸根释放出 1.5 mol 原子氧：

$$2KMnO_4 + H_2O \longrightarrow 2MnO_2\downarrow + 2OH + 3[O]$$

在强酸性介质中，锰原子的价态则由+7 降至+2，形成二价锰盐，平均 1 mol 高锰酸根释放出 2.5 mol 原子氧：

$$2KMnO_4 + 3H_2SO_4 \longrightarrow 2MnSO_4\downarrow + K_2SO_4 + 3H_2O + 5[O]$$

例如，在碱性条件下，如果将 1 mol 乙苯氧化为苯甲酸，则需要 4 mol 高锰酸钾：

相对来说，高锰酸钾在中性介质中的氧化反应要温和一些，适用于由烯烃制备邻二醇。

$$C_6H_5CH_2CH_3 + 6[O] \longrightarrow C_6H_5COOH + CO_2 + 2H_2O$$

在碱性介质中，高锰酸钾可以将伯醇或醛氧化为相应的酸，也可以用来氧化芳烃上的侧链。

$$CH_3(CH_2)_7CH{=}CH(CH_2)_7COOH \xrightarrow{KMnO_4,\ H_2O} CH_3(CH_2)_7CH(OH)CH(OH)(CH_2)_7COOH$$

$$CH_3CH(CH_3)CH_2OH \xrightarrow{KMnO_4,\ OH^-} CH_3CH(CH_3)COOH$$

$$p\text{-}CH_3C_6H_4COCH_3 \xrightarrow{KMnO_4,\ OH^-} p\text{-}HOOCC_6H_4COOH$$

高锰酸钾在酸性介质中的氧化反应常在 25%硫酸溶液或更浓的硫酸溶液中进行。在酸性介质中，高锰酸钾对烷基芳烃的氧化常伴随着脱羧反应，因而酸性介质法用得很少。

如何确定待氧化反应物与 $KMnO_4$ 的配比，以甲苯在碱性条件下氧化成苯甲酸为例，每氧化 1 mol 甲苯需消耗 3 mol 原子氧，也就是说，1 mol 甲苯需投入 2 mol 的 $KMnO_4$：

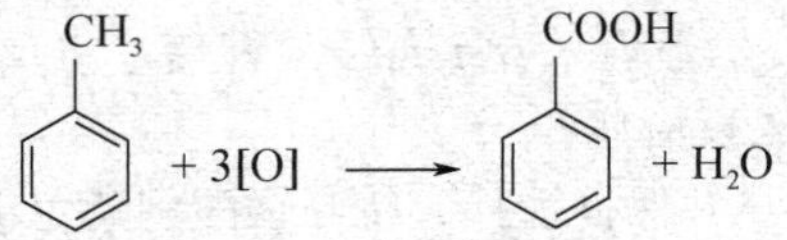

$$C_6H_5CH_3 + 3[O] \longrightarrow C_6H_5COOH + H_2O$$

除了丙酮、醋酸、叔丁醇等少数溶剂外，高锰酸钾一般不溶于有机溶剂，其氧化反应多在水溶液中进行。以高锰酸钾水溶液对有机化合物进行氧化，反应发生在水相与有机相之间，反应速度比较慢，收率也比较低。如果加入一些相转移催化剂，则会产生奇效。

硝酸作为氧化剂，1 mol 硝酸可释放出 1.5 mol 原子氧：

$$2HNO_3 \longrightarrow 2NO_2 + H_2O + 3[O]$$

由于硝酸氧化性很强，通常用于羧酸的制备，例如由环己醇经硝酸氧化生成己二酸。

3.5.2　高锰酸钾氧化实验通法

高锰酸钾氧化法多在碱性条件下应用，现以烷基芳烃氧化为芳烃羧酸为例，按 0.1 mol 烷基芳烃投入量计，依次将 0.1 mol 烷基芳烃、0.1 mol 碳酸钠和 600 mL 水加入到 1000 mL 三口烧瓶中，搅拌下将 0.3 mol 高锰酸钾（高锰酸钾的投入量是以芳烃上一个甲基形成一个羧基的关系而计量的）分次投入到反应瓶中。加热回流，每加入一份高锰酸钾都要待反应液紫色褪尽后再加另一批。加料完毕，如果紫色很快褪去，需再补加适量的高锰酸钾（约占总投入量的 10%），然后，继续搅拌回流 2～4 h。

趁热过滤，用沸水将二氧化锰沉淀洗涤 3 次。合并滤液，用 50%硫酸进行酸化，析出芳羟羧酸沉淀，过滤，并用少量冷水洗涤，再用水重结晶。

3.5.3　实验

实验六　烟酸（抗糙皮病药物）

【实验目的】

学习高锰酸钾氧化法对烷基芳烃的氧化原理及实验方法，学习水溶性有机化合物分离纯化技术。

【实验原理】

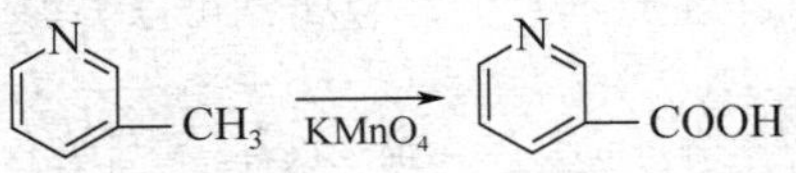

$$\text{3-甲基吡啶}\ (C_5H_4N{-}CH_3) \xrightarrow{KMnO_4} \text{烟酸}\ (C_5H_4N{-}COOH)$$

烟酸为无色针状结晶，mp236～239℃。

【药品】

3-甲基吡啶	3.0 g（3.1 mL，0.032 mol）
高锰酸钾	12 g（0.075 mol）
浓盐酸	12～13 mL

【实验操作】

在 250 mL 三口烧瓶上，配置搅拌器、粉末固体漏斗和温度计。将 3 g 3-甲基吡啶和 100 mL 水加入到三口烧瓶中，三口烧瓶置于水浴中加热至 70℃。在搅拌下，将 12 g 高锰酸钾分成 10 份分批投料。每加入一批高锰酸钾后，要待反应液紫红色褪去后再加入下一批。最初投料时反应温度保持在70℃，当投入6 g高锰酸钾后，将反应温度提高至85～90℃，再将剩余的6 g高锰酸钾分批投入反应瓶。

加料完毕，在沸水浴上加热并保持搅拌。待高锰酸钾紫色褪尽后趁热过滤，用热水将二氧化锰滤饼洗3～4次（每次10 mL），合并滤液于烧杯中，加热浓缩滤液至100 mL左右。然后用滴管向浓缩液滴加浓盐酸（约4 mL），将溶液的pH值调至3.4（烟酸的等电点）。

注意：用精密pH试纸检测。

将溶液静置冷却（或置于冰箱中过夜），使烟酸晶体慢慢析出。

过滤、收集固体产物并用少量冷水洗涤，抽滤后置粗产物于90～100℃条件下干燥。将滤液蒸发浓缩至60 mL，然后慢慢冷却至5℃，又可得第二批产物。

粗产物可用水重结晶。

【注意事项】

慢慢冷却结晶，有利于减少氯化钾在产物中的夹杂量。

【思考题】

（1）烟酸在水中的溶解度（g/100 mL）数据如下：

0.10（0℃）、0.26（40℃）、0.82（80℃）、1.27（100℃）

试拟定对2 g烟酸粗品重结晶的实验方案。

（2）在产物后处理过程中，为什么要将pH值调至烟酸的等电点？

（3）本实验在对反应混合物后处理过程中，为什么强调对第二次浓缩液要作慢速冷却结晶处理？冷却速度过快会造成什么后果？

（4）如果在烟酸产物中含有少量氯化钾，如何除去？试拟定分离纯化方案。

3.6 还原反应

在有机合成化学中，能使有机分子增加氢原子或减少氧原子的一类反应称为还原反应（Reduction Reaction），还原反应在精细有机合成中占有重要的地位。常用的还原方法有金属与供质子剂（如酸、醇等）还原、催化氢化、金属氢化物还原等。其中，金属与供质子剂还原法在实验室中应用较为广泛。例如锂、钠、钾、镁、锌、锡、铁等金属，它们的电动势均大于氢，都可以用作还原剂。常用的供质子剂有酸、醇、水、氨等。

如果金属与供质子剂反应太强烈，则还原效果不好，因为质子会以分子氢的形式逸出。例如金属钠与盐酸就不可用来作还原剂，但是金属钠可以与醇一起作还原剂。

$$H_3C\text{—}C(=O)\text{—}C_5H_{11} \xrightarrow{Na+CH_3CH_2OH} CH_3CH(OH)C_5H_{11}$$

电动势大于氢的各种金属与不同的供质子剂组合在一起，数量很多，而且它们的还原性能还与反应条件以及被还原物结构有密切的关系。因此，在这里不便以一个实验通法来概括全貌。

3.6.1 羰基还原实验通法（黄鸣龙还原法）

在500 mL三口烧瓶上，配置温度计、搅拌器和冷凝管（见图2-6），依次向反应烧瓶加入0.1 mol羰基化合物、13 g（0.23 mol）氢氧化钾及180 mL二甘醇。开启搅拌，加热至130℃，在此温度下搅拌反应2 h。然后将反应装置改为蒸馏装置，蒸出多余的肼和水，直至反应混合物的温度升到195～210℃，再将蒸馏装置改为回流装置，并维持回流，直到无氮气放出为止（需时约4 h）。冷却后，向反应混合物中加入等体积水稀释，然后，用乙醚提取3次。乙醚提取液依次用稀盐酸、水、5%Na_2CO_3水溶液、水进行洗涤，经无水硫酸镁干燥，蒸除溶剂。最后，对残余物进行蒸馏或重结晶。

3.6.2　硝基芳烃铁屑还原实验通法

通常，1 mol硝基化合物的还原需要3～4 mol的铁屑，大大超过理论值。在以铁屑作还原剂的反应过程中，电解质的存在可提高溶液的导电能力，加速铁的腐蚀过程，加快还原速度。研究表明，在铁屑对硝基苯的还原反应中，不同的电解质对还原速度影响的活性顺序为：

$$NH_4Cl > FeCl_2 > (NH_4)_2SO_4 > BaCl_2 > CaCl_2 > NaCl > Na_2SO_4 > KBr > NaAc$$

水在铁屑还原硝基芳烃的反应中，既作介质又作还原反应中的氢源。水与硝基芳烃的用量比为50～100∶1。对于低活性硝基芳烃，可以加入甲醇、乙醇等与水相混，有利于反应。

在圆底烧瓶上配置回流冷凝管，依次加入0.3 mol铁粉、2 g氯化铵及50 mL水。边搅拌边加热，小火沸煮15 min。稍后入0.1 mol硝基芳烃，搅拌并回流1.5 h。冷却至室温后进行处理。根据胺类产物的不同性质，可以采用相应的分离提纯方法。

（1）对于不溶于水且具有一定蒸气压的芳胺，可以采用水蒸气蒸馏法分离。例如苯胺、对甲苯胺、邻甲苯胺、对氯苯胺、邻氯苯胺等。

（2）对于易溶于水且可蒸馏的芳胺，可以采用过滤除铁粉、简单蒸馏除水分、最后作减压蒸馏的方法分离。例如间苯二胺、对苯二胺、2,4-二氨基甲苯等。

（3）对于易溶于热水的芳胺，可以采用先热过滤、然后冷却结晶的方法分离。例如邻苯二胺、邻氨基苯酚、对氨基苯酚等。

（4）对于不溶于水且蒸气压很低的芳胺，可以采用溶剂萃取的方法提取。例如α-萘胺。

【注意事项】

氯化铵水解后生成盐酸，铁粉经稀盐酸处理后，可以提高反应活性。

3.7　威廉森反应

3.7.1　威廉森反应实验原理

醚的制备方法有多种，如醇脱水、硫酸二烷基酯和酚盐作用或威廉森反应等。以醇脱水的制醚法常用于制取单醚（也称对称醚），如甲醚、乙醚等。若用两种不同的醇经脱水制混合醚（也称不对称醚），则会生成好几种醚的混合物，分离较困难。除了制取芳基烷基醚外，一般很少用脱水法制备混合醚。由于硫酸二甲酯、硫酸二乙酯等烷基化试剂毒性很大，因而采用威廉森反应制备混合醚最为常见，它是以卤代烃和醇钠经亲核取代反应来制取醚。

$$RX + R'OM \longrightarrow ROR' + MX$$

在威廉森反应中，卤代烃的反应活性的顺序是：RI＞RBr＞RCl。其中，碘代烷的反应活性最高。如果在丙酮或醇溶液中用RBr或RCl进行反应时，加入10%（相对于卤代烃投入量）的碘化钾或碘化钠常能加快反应。另外，卤代烃分子中的烷基大小对于反应走向也有明显的影响。通常，随着烷基上的支链增多，生成烯烃的倾向也就越大。例如，以叔卤代烷与醇钠作用，主要产物是烯烃而不是醚。

$$H_3C-C(CH_3)_2-Br + NaOCH_3 \longrightarrow (H_3C)_2C{=}CH_2 + CH_3OH + NaX$$

因此，在具有仲或叔烷基混醚的合成中，仲或叔烷基常以醇盐而不是卤代烃的形式引入。仍以叔丁基醚的制备为例。

$$H_3C-C(CH_3)_2-ONa + ICH_3 \longrightarrow H_3C-C(CH_3)_2-OCH_3 + NaI$$

威廉森反应既可用于制备分子量较大的烷基类单醚和混合醚，也可用于合成芳基烷基醚。在含有芳烃的混合醚制备中，由于卤代芳烃分子中的卤代原子不活泼，不易发生取代反应，而酚的酸性比醇强，易生成盐。因此，应采用酚钠和卤代烃反应的途径来合成。

$$C_6H_5ONa + CH_3I \longrightarrow C_6H_5OCH_3 + NaI$$

由于酚的酸性较强，在芳基烷基混合醚的制备中，可以用酚和苛性碱制取酚盐继而进行威廉森反应。

$$C_6H_5OH + NaOH \longrightarrow C_6H_5ONa + H_2O$$

3.7.2 威廉森反应实验通法

在三口烧瓶上配置滴液漏斗和连有氯化钙干燥管的回流冷凝管。向烧瓶中加入 2.0 g（0.13 mol）切成片状的金属钠，然后滴加 1 mol 干燥醇。醇的滴入速度以保持溶液平稳沸腾为宜。

注意：废弃的金属钠屑不要弃在水槽中，以防意外。可以将钠屑浸泡在异丙醇中加以处理。有金属钠参与的反应切不可用水浴加热。

当金属钠完全溶解后，自滴液漏斗向三口烧瓶中滴加 0.1 mol 卤代烃（或者 0.1 mol 硫酸二烷基酯），油浴加热回流 2 h，以促进反应进行。

注意：如果使用的卤代烃反应活性太低，如溴代烃，可在滴加卤代烃的同时加入 1 g 左右无水碘化钾。

如果制备芳醚，如苯基醚，可以依下法制备：在圆底烧瓶中加入 4.4 g（0.11 mol）氢氧化钠和 30 mL 水。待氢氧化钠完全溶解后，加入 9.4 g（0.1 mol）苯酚，迅速搅拌使之溶解。然后滴加 0.1 mol 卤代烃，加热回流 2 h。待反应混合物冷却后倾入 50 mL 水中，再用乙醚萃取 3 次，合并有机相，经水洗涤后用氯化钙干燥。过滤、蒸除乙醚，再对残余物进行蒸馏或减压蒸馏或重结晶。

如果需要对未反应的酚进行回收，可依下法处理：用稀盐酸对经萃取后的碱性水溶液进行酸化，再用乙醚萃取，经洗涤、干燥后，蒸除乙醚，即可回收未反应的酚。

3.7.3　实验

实验七　甲基叔丁基醚（无铅汽油抗震剂）

【实验目的】

学习威廉森制醚法原理及实验方法。

【实验原理】

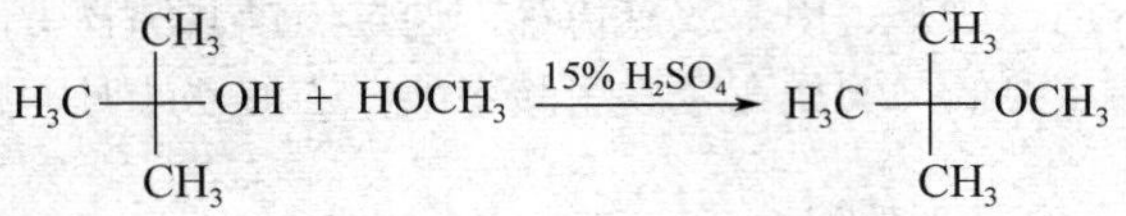

$$H_3C-\underset{CH_3}{\overset{CH_3}{\underset{|}{\overset{|}{C}}}}-OH + HOCH_3 \xrightarrow{15\%\ H_2SO_4} H_3C-\underset{CH_3}{\overset{CH_3}{\underset{|}{\overset{|}{C}}}}-OCH_3$$

甲基叔丁基醚为无色透明液体，bp55～56℃，n_D^{20}1.3690，d0.740。

【药品】

叔丁醇	14.8 g（19 mL，0.2 mol）
甲醇	12.8 g（16 mL，0.4 mol）
15%硫酸	70 mL
无水碳酸钠	3～5 g

【实验操作】

在 250 mL 圆底烧瓶上配置分馏柱，分馏柱顶端装上温度计，在其支管处依序配置直形冷凝管、接引管和接收瓶。接引管支管连接橡皮管并导入水槽。接收瓶置于冰浴中。

将 70 mL 5%硫酸、16 mL 甲醇和 19 mL 叔丁醇加入到圆底烧瓶中，振摇使之混合均匀。投入几颗沸石，小火加热。收集 49～53℃时的馏分。

将收集液转入分液漏斗，依次用水、10%Na_2SO_3水溶液、水洗涤，以除去醚层中的醇和可能存在的过氧化物。当醇洗净时，醚层显得清澈透明。然后，用无水碳酸钠干燥蒸馏、收集 53～56℃时的馏分。称量、测折射率并计算产率。

【注意事项】

叔丁醇熔点为 25.5℃，沸点为 82.5℃，有少量水存在时呈液体。如果室温较低，加料困难时，可以加入少量水，使之液化后再加料。

【思考题】

（1）通常，混合醚的制备宜采用威廉森合成法，为什么本实验可以用硫酸催化脱水法制备混合醚甲基叔丁基醚？

（2）为什么要以稀硫酸作催化剂？如果采用浓硫酸会使反应产生什么结果？

（3）反应过程中，为何要严格控制馏出温度？馏出速度过快或馏出温度过高，会对反应带来什么影响？

3.8　酯化反应

羧酸与醇或酚在无机或有机强酸催化下发生反应生成酯和水，这个过程称为酯化反应。常用的催化剂有浓硫酸、干燥的氯化氢、有机强酸或阳离子交换树脂。在酯化反应中，如果参与反应的羧酸本身就具有足够强的酸性，例如甲酸、草酸等，那就可以不另加催化剂。

酯化反应是一个可逆反应，当酯化反应达到平衡时，通常只有65%左右的酸和醇反应生成酯。

$$RCOOH+R'OH \xrightleftharpoons{H^+} RCOOR'+H_2O$$

为了使反应向有利于酯的生成的方向进行，可以从反应物中不断移去产物酯或水，或者使用过量的羧酸或醇。至于究竟是用过量的酸还是过量的醇，这就取决于原料的性质及价格等因素。例如，在合成乙酸乙酯时，由于乙醇比乙酸便宜，因而加入过量的乙醇与乙酸反应。另外，为了除去反应中生成的水，通常采用共沸蒸馏法，即在酯化反应混合物中加入一些能与水共沸的有机溶剂，如苯、甲苯或氯仿等，通过蒸馏共沸物带出生成的水。如果酯的沸点比酸、醇及水的沸点要低，例如在合成甲酸甲酯、乙酸乙酯时，则可采取不断蒸除酯的方法使平衡正向移动。

3.8.1 酯化反应实验通法

在三口烧瓶上配置油水分离器（如果酯的沸点比原料及水的沸点低，就不必安装油水分离器，可以直接采用蒸馏装置，边反应边将产物酯蒸出）和回流冷凝管。依次向反应瓶中加入0.1 mol羧酸、0.12 mol醇（或者使用过量的羧酸）、50 mL苯和1 mL浓硫酸。投入几粒沸石，加热回流，蒸气经冷凝管冷凝后流入到油水分离器中，冷凝液主要组分为苯和水。当上层的苯层积聚至油水分离器支管处时，会不断流回到反应瓶；当下层的水相积聚较多时，可以打开油水分离器活塞放出水层。

反应结束后，将反应混合物移至分液漏斗中，加入40 mL水，振摇后分除下面水层。留在分液漏斗中的酯层，依次用20 mL水、10 mL 5%碳酸钠水溶液洗涤，然后再用水洗涤数次，使有机相呈中性，用无水硫酸镁干燥后蒸馏。

【注意事项】

（1）如果产物酯自身能与水形成共沸物，也可以不再另加苯或其他有机溶剂。

（2）根据醇及酯的性质，苯层中也可能含有不同比例的醇和酯。

（3）可根据收集的水的体积来判断反应终点，不过，水的实际收集量要比理论计算值高；因为水常以共沸物形式蒸出，故应以等量水所形成的相应共沸物体积来判断反应终点。也可通过观察油水分离器中是否有水珠继续下沉来判断终点。

（4）如果水层和酯层分层困难，可以加入饱和食盐水洗涤。

3.8.2 实验

实验八　乙酰水杨酸（阿司匹林）

【实验目的】

学习以酚类化合物作原料制备酯的原理和实验方法，巩固重结晶操作技术。

【实验原理】

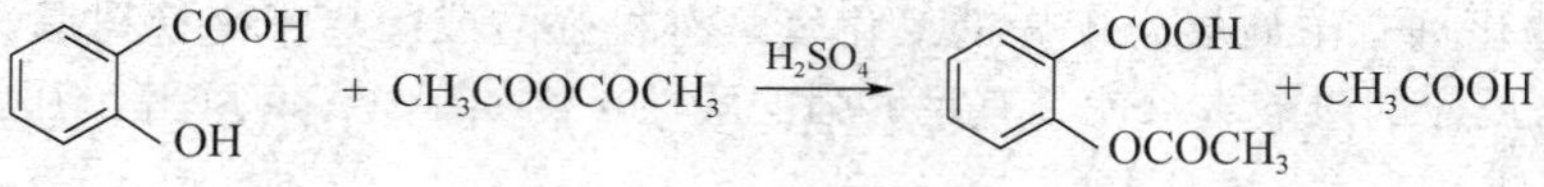

$$\text{C}_6\text{H}_4(\text{COOH})(\text{OH}) + CH_3COOCOCH_3 \xrightarrow{H_2SO_4} \text{C}_6\text{H}_4(\text{COOH})(\text{OCOCH}_3) + CH_3COOH$$

乙酰水杨酸为白色针状晶体，mp132～135℃（乙酰水杨酸受热易分解，熔点不明显。测定时，可先将浴液加热至 110℃左右，再将待测样品置入其中测定）。

【药品】

水杨酸	1.38 g（0.01 mol）
乙酸酐	4 mL（0.04 mol）
浓硫酸	少量
10%碳酸氢钠水溶液	20 mL
20%盐酸	10 mL
1%三氯化铁溶液	少量

【实验操作】

在 100 mL 锥形瓶中依次加入 1.38 g 水杨酸、4 mL 乙酸酐和 4 滴浓硫酸，摇匀，使水杨酸溶解。

注意：乙酸酐和浓硫酸均具有强腐蚀性，量取时要当心。若不慎溅及皮肤，立即用大量水冲洗。

将锥形瓶置于 60～70℃的热水浴中，加热 10 min，并不时地振摇。然后，停止加热，待反应混合物冷却至室温后，缓缓加入 15 mL 水，边加水边振摇。将锥形瓶放在冷水浴中冷却，有晶体析出。抽滤，并用少量冷水洗涤，抽干，得乙酰水杨酸粗产品。

注意：由于剩余的乙酸酐发生水解，反应瓶会变热，有时反应混合甚至会沸腾，操作要当心。

将粗产品转入到 100 mL 烧杯中，加入 10%碳酸氢钠水溶液，边加边搅拌，直到不再有二氧化碳产生为止。抽滤，除去不溶性聚合物。再将滤液倒入 100 mL 烧杯中，缓缓加入 10 mL 20%盐酸，边加边搅拌，这时会有晶体逐渐析出。将反应混合物置于冰水浴中，使晶体尽量析出。抽滤，用少量冷水洗涤 2～3 次，然后抽滤至干。取少量乙酰水杨酸，溶入几滴乙醇中，并滴加 1～2 滴 1%三氯化铁溶液，如果发生显色反应，产物可用乙醇-水混合溶剂重结晶：先将粗产品溶于少量的沸乙醇中，再向乙醇溶液中添加热水直到溶液中出现混浊，再加热至溶液澄清透明，静置慢慢冷却，过滤、干燥、称量，测定熔点并计算产率。

【注意事项】

（1）酚类化合物的酯化也称酰化，常用的酰化试剂有酰氯、酸酐等。与酰氯相比，酸酐和酚类化合物的反应要温和一些。

（2）添加少量浓硫酸或浓磷酸会加速反应。

（3）在反应过程中，少量水杨酸自身会发生聚合反应，形成一种聚合物。阿司匹林可以与碳酸氢钠作用形成水溶性盐，从而与聚合物分离。

（4）酚类化合物能与三氯化铁溶液发生显色反应，这种特殊的显色反应可用来检验酚羟基的存在。

（5）在乙酰水杨酸重结晶时，其溶液不宜加热过久，且不宜用高沸点溶剂，因为在高温下乙酰水杨酸易发生分解。

【思考题】

（1）在水杨酸与乙酸酐的反应过程中，浓硫酸起什么作用？

（2）纯的乙酰水杨酸不会与三氯化铁溶液发生显色反应。然而，在乙醇-水混合溶剂中经过重结晶的乙酰水杨酸，有时反而会与三氯化铁溶液发生显色反应，这是什么缘故？

（3）水杨酸与乙酸酐的反应结束后，如果不采用碳酸氢钠成盐、盐酸酸化的方法分离聚合物杂质，你可否另拟定一个分离纯化的方案？

Section Ⅲ

General Method for Organic Compound Synthesis

3.1 Halogenation Reactions

3.1.1 Principles of Halogenation Reactions

Introducing halogen atoms into organic molecules to prepare halogenated hydrocarbons is called halogenation. Halogenated hydrocarbons are not only important intermediates in organic synthesis but also commonly used organic solvents. Many useful compounds, such as alcohols, phenols, ethers, amines, aldehydes, ketones, and acids, can be derived from halogenated hydrocarbons. Therefore, halogenation reactions are widely applied in organic synthesis.

The methods and reagents used for the preparation of halogenated hydrocarbons differ depending on the hydrocarbon structure in the molecule. For example, alkyl halides can be obtained by reacting hydrocarbons with hydrogen halides or by free radical substitution reactions of hydrocarbons with halogen molecules under light conditions. Aryl halides can be obtained by electrophilic substitution reactions of aromatic hydrocarbons with chlorine or bromine in the presence of iron powder or the corresponding iron (Ⅲ) halide catalysts.

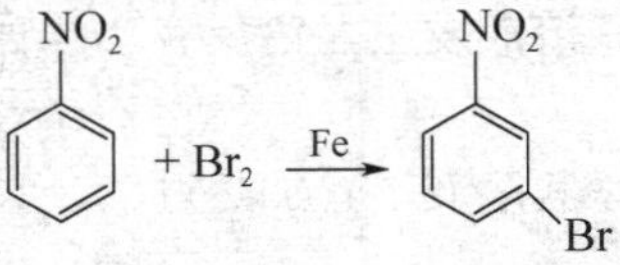

$$NO_2 \qquad \qquad NO_2$$
$$+ Br_2 \xrightarrow{Fe}$$
$$Br$$

Direct reaction between fluorine and aromatic hydrocarbons is too vigorous and may even lead to decomposition of the aromatic compound. Therefore, fluori nated aromatic compounds are usually prepared from the corresponding aniline through diazotization and Sandmeyer reactions. Although iodination of aromatic compounds can be achieved by direct reaction with iodine, the presence of iodine hydride produced simultaneously reduces the reaction reversibility. For example:

$$ArH + I_2 \rightleftharpoons ArI + HI$$

Obviously, the addition of oxidants such as concentrated sulfuric acid or nitric acid to the reaction system can promote the forward iodination reaction. If halogenated hydrocarbons are used as synthetic intermediates, bromination is the most suitable choice during the laboratory preparation of halogenated hydrocarbons because bromine is a liquid at room temperature and easy to handle.

3.1.2 General Procedure for Bromination of Aromatic Hydrocarbons

For a feeding amount of 0.1 mol of aromatic hydrocarbon, a 250 mL three-neck flask can be used as the reaction vessel, equipped with a stirrer, a reflux condenser, a dropping funnel, a thermometer, and a hydrogen bromide gas absorption device（see Figure 2-6, Figure 2-7）.

The bromination conditions of aromatic hydrocarbons are closely related to their reactivity.

If the reactivity of the aromatic hydrocarbon is low, add 0.1 mol of aromatic hydrocarbon and 0.5 g of iron powder to the three-neck flask, and heat it in an oil bath to 100～150°C. Then, while stirring, slowly add 10 g（0.06 mol）of bromine through the dropping funnel into the three-neck flask at a rate that prevents bromine vapor from escaping through the condenser. After the addition is complete, continue stirring and heating the reaction for 1 hour, then add 0.5 g of iron powder and drop 10 g of bromine. After the addition is complete, continue stirring and heating the reaction for 2 hours.

Pour the reddish-brown reaction mixture containing a small amount of bromine into a 200 mL aqueous solution pre-dissolved with 1 g of sodium bisulfite to remove the residual bromine. If the bromine color has not been completely removed, a small amount of sodium bisulfite can be added. Usually, there will be some unreacted bromine remaining in the bromination product, which is not easily washed away with water. However, using sodium bisulfite can reduce it to water-soluble sodium bromide and remove it:

$$Br_2 + 3NaHSO_3 \longrightarrow 2NaBr + NaHSO_4 + 2SO_2 + H_2O$$

Then, perform steam distillation on the mixture. If the product is a solid, it can be filtered, air-dried, and meet the general requirements for use. If necessary, recrystallization can be carried out. If the product is a liquid, it can be extracted with carbon tetrachloride. If the product is a liquid insoluble in water, it can be separated directly using a separatory funnel. Then, wash the product successively with 10% sodium hydroxide solution and water until neutral, dry with magnesium sulfate, and remove the solvent by evaporation under reduced pressure.

When the reactivity of the aromatic hydrocarbon is moderate, add 0.1 mol of aromatic hydrocarbon and 0.2 g of iron powder to the three-neck flask, then slowly add 16 g（0.1 mol）of bromine while stirring at room temperature. After the addition of bromine, stir the reaction for 1 hour at room temperature, during which hydrogen bromide gas is evolved. If the reaction is too slow, heat the reaction for 1 hour using a water bath（30～40°C）. Then, increase the water bath temperature to around 65°C and continue stirring the reaction for some time until there is no more reddish-brown vapor escaping from the reaction mixture. The subsequent treatment is the same as before.

When the reactivity of the aromatic hydrocarbon is high, bromination must be carried out under mild conditions. Dilute 0.1 mol of aromatic hydrocarbon with 50 mL of carbon tetrachloride and cool it to 0°C, then slowly add a 0.08 mol bromine solution in carbon tetrachloride (prepared from 10 mL of carbon tetrachloride and 13 g of bromine) while stirring. Control the reaction temperature at 0-5°C. After the addition of bromine, continue stirring at low temperature (0～5°C) for 1～2 hours until there is no more reddish-brown vapor escaping from the reaction mixture. The subsequent treatment is the same as before.

It is important to note that bromine is highly corrosive and can cause severe burns to the skin. Its vapors have an irritating effect on mucous membranes. Therefore, it must be handled with rubber gloves in a fume hood during measurement.

When measuring the bromine, place the separatory funnel placed on an iron ring inside the fume hood, then pour the bromine in.

3.1.3 Experiment

Experiment One *p*-Bromoacetanilide

【Experimental objectives】

To learn the theory of aromatic halogenation reactions, master the bromination method of aromatic compounds, become familiar with the physical and chemical properties of bromine and its handling methods, and acquire the techniques of recrystallization and melting point determination.

【Experimental principle】

$$C_6H_5NHCOCH_3 \xrightarrow{Br_2/HAc} p\text{-}BrC_6H_4NHCOCH_3$$

【Chemicals】

Acetanilide	13.5 g (0.1 mol)
Bromine	16 g (5 mL, 0.1 mol)
Glacial acetic acid	36 mL
Sodium bisulfite	1～2 g

【Experimental procedure】

Set up a stirring apparatus, thermometer, dropping funnel, and reflux condenser with a gas absorption device attached to absorb the generated hydrogen bromide during the reaction using a 250 mL three-neck flask. Note: Ensure good sealing between the stirring apparatus and the neck of the flask to prevent hydrogen bromide from overflowing.

Add 13.5 g of acetanilide and 30 mL of glacial acetic acid to the three-neck flask. Slightly heat the mixture in a warm water bath to dissolve the acetanilide. Then, under a bath temperature of 45℃, add dropwise a solution prepared by mixing 16 g of bromine and 6 mL of glacial acetic acid while stirring. The addition speed should be such that the reddish-brown color of bromine fades relatively quickly. Note: Bromine is highly corrosive and irritating. It must be handled in a fume hood. Wear rubber gloves during the operation.

After completing the addition, continue stirring the reaction mixture at 45℃ for 1 hour. Then raise the bath temperature to 60℃ and stir for a period of time until no more reddish-brown vapors are released from the reaction mixture.

Pour the reaction mixture into a beaker containing 200 mL of cold water（if the product has a reddish-brown color, dissolve 1 g of sodium bisulfite in cold water beforehand; if the color of the product is still dark, add a suitable amount of sodium bisulfite）. Stir the mixture with a glass rod for 10 minutes, then filter it after the mixture cools to room temperature. Wash the filtered cake with cold water and suction dry it. Dry the product at a temperature of 50～60℃. The resulting *p*-bromoacetanilide can be used directly for the preparation of *p*-bromoaniline.

p-Bromoacetanilide can be recrystallized using methanol or ethanol. After drying the product, weigh it, measure the melting point, and calculate the yield. *p*-Bromoacetanilide is a colorless crystals with a melting point range of 164～166℃.

【Notes】

（1）When the room temperature is below 16℃, glacial acetic acid solidifies. It can be melted by placing the reagent bottle containing glacial acetic acid in a warm water bath.

（2）The dropping rate should not be too fast. Otherwise, the reaction will be too vigorous, causing some bromine to escape without participating in the reaction and leading to the formation of dibromo derivatives.

【Thought questions】

（1）Why is the para-isomer the major product of the bromination of acetanilide?

（2）How does the reaction temperature affect the outcome of the bromination reaction?

（3）What is the purpose of washing the reaction mixture with a sodium bisulfite aqueous solution in the post-treatment process?

（4）What impurities might be present in the product, and how can they be removed?

3.2 Sulfonation Reaction

3.2.1 Principle of Sulfonation Reaction

The reaction in which hydrogen atoms in organic molecules are replaced by sulfonic acid groups（—SO_3H）is called the sulfonation reaction. Sulfonation of aromatic hydrocarbons is an electrophilic substitution reaction and is a widely used unit reaction. Introducing sulfonic acid groups onto aromatic rings can enhance water solubility, which is of great significance in the dye and pharmaceutical synthesis industry. Sulfonic acid compounds are important intermediates in organic synthesis because sulfonic acid groups can be easily converted into hydroxyl, amino, nitro, cyanide, etc. Furthermore, many sulfonated products of aromatic hydrocarbons themselves have important applications. For example, alkylbenzene sulfonates with 12～15 carbon atoms in the alkyl group can be used as detergents. Lower alkyl naphthalene sulfonates are widely used wetting agents and emulsifiers, such as sodium dodecyl naphthalene sulfonate. Commonly used sulfonating agents include concentrated sulfuric acid, chlorosulfonic acid（$ClSO_3H$）, sulfur trioxide, etc. Among them, sulfuric acid is the mildest sulfonating agent and is usually used for sulfonating more reactive aromatic hydrocarbons. Chlorosulfonic acid is a more vigorous sulfonating agent; it can not only sulfonate aromatic hydrocarbons but also sulfonate aliphatic hydrocarbons. When aromatic hydrocarbons react with equimolar amounts of chlorosulfonic acid, aryl sulfonic acids are formed. If excess chlorosulfonic acid is used, aryl sulfonyl chlorides are formed. Sulfur trioxide is the strongest sulfonating agent. Smoke sulfuric acid（sulfur trioxide dissolved in sulfuric acid）can be used to sulfonate less reactive aromatic hydrocarbons. However, using sulfur trioxide as a sulfonating agent can easily result in oxidation reactions. Therefore, it is advisable to carry out the sulfonation reaction at lower temperatures or use haloalkanes as diluents to moderate the reaction.

Sulfonation product: Similar to sulfuric acid, sulfonic acid is a highly soluble strong acid that can dissolve in excess sulfonating agents. After the sulfonation reaction is completed, the reaction mixture is usually diluted with ice water first,

NH_2 (benzene ring) $+ H_2SO_4 \xrightarrow{185℃}$ NH_2 (benzene ring) SO_3H

NO_2 (benzene ring) $+ 2ClSO_3H \xrightarrow{105℃}$ NO_2 (benzene ring) SO_2Cl

then neutralized with a base and saturated salt water is added to precipitate the sulfonic acid as a salt（salt precipitation method）:

$$ArSO_3H + NaCl \rightleftharpoons ArSO_3Na\downarrow + HCl$$

Unlike sulfuric acid, calcium and barium salts of sulfonic acid are soluble in water. This difference can be used to neutralize them with calcium carbonate（barium）and filter out the precipitated sulfate salts to remove excess sulfuric acid. The filtrate can then be treated with a solution of sodium carbonate to remove the generated calcium carbonate precipitate, resulting in sodium sulfonate salt（this purification method of sodium sulfonate salt is also called desulfation by calcium carbonate）.

$$\left.\begin{matrix} H_2SO_4 \\ ArSO_3H \end{matrix}\right\} + CaCO_3 \longrightarrow (ArSO_3)_2Ca + CaSO_4$$

Using chlorosulfonic acid as a sulfonating agent, the sulfonyl chloride product is slightly soluble in water, making separation and purification easier. Many sulfonyl chlorides can be purified by distillation, and sulfonic acid can be obtained by hydrolysis of sulfonyl chlorides. In addition, due to the high reactivity of sulfonyl chlorides, many useful sulfonic acid derivatives such as sulfonamides and esters can be prepared from them. Therefore, in the laboratory, chlorosulfonic acid is more commonly used.

3.2.2 General Method for Chlorosulfonation of Aromatic Hydrocarbons

For a feed amount of 0.1 mol of aromatic hydrocarbon, a 250 mL three-necked flask can be used as the reaction vessel, equipped with a stirrer, reflux condenser, dropping funnel, thermometer, and gas absorption device（see Figure 2-6 and Figure 2-7）. The reaction conditions of aromatic hydrocarbon chlorosulfonation are closely related to the reactivity of the aromatic hydrocarbon. If the reactivity of the aromatic hydrocarbon is low, 0.3 mol of chlorosulfonic acid and 0.1 mol of aromatic hydrocarbon can be added to the three-necked flask, heated and stirred, and slowly heated to 110～120℃. At this time, a large amount of hydrogen chloride will evolve. When the evolution of hydrogen chloride gas slows down, indicating that the reaction is nearing completion, the reaction temperature can be increased by 10℃ and stirring continues until no more hydrogen chloride gas is evolved. If the reactivity of the aromatic hydrocarbon is high, in order to avoid an overly vigorous reaction, 0.1 mol of aromatic hydrocarbon can be diluted with 25 mL of dried chloroform and cooled to around −10℃ using an ice-salt bath. Under vigorous stirring, 0.2 mol of chlorosulfonic acid is slowly added dropwise. At this time, a large amount of hydrogen chloride gas will be evolved. After the addition is complete, stirring should continue at around −10℃. When the evolution of hydrogen chloride gas slows down, the reaction mixture can be gradually warmed to room temperature and stirring continued until no more hydrogen chloride is evolved. If the reactivity of the aromatic hydrocarbon is moderate, 0.1 mol of aromatic hydrocarbon can be added to the flask and cooled to 0℃. Under vigorous stirring, 0.25 mol of chlorosulfonic acid is added dropwise. After the addition is complete, stirring should continue at room temperature until no more hydrogen chloride gas evolves. After the reaction is completed, the reaction mixture is slowly poured into 100 g of crushed ice while stirring in a fume hood. If the sulfonic chloride precipitate is a solid product, filtration, washing, and recrystallization can be performed；if it is a liquid product, the reaction mixture can be extracted with solvents such as chloroform or benzene, followed by washing with water, a solution of sodium bicarbonate, and water, and finally subjected to distillation to obtain the chlorosulfonated product. Note: Chlorosulfonic acid is similar to concentrated sulfuric acid in that it is highly acidic and corrosive and can cause skin burns. Rubber gloves should be worn when handling and operations should be conducted in a fume hood.

3.2.3 General Method for Sulfonation of Aromatic Hydrocarbons

For a feed amount of 0.1 mol of aromatic hydrocarbon, a 100 mL three-necked flask can be used as the reaction vessel, equipped with a stirrer, reflux condenser, and thermometer（see Figure 2-6）. If the reactivity of the aromatic hydrocarbon is low, the three-necked flask can be placed in an ice-water bath, 35 g of 25% fuming sulfuric acid can be added, and 0.1 mol of aromatic hydrocarbon can be slowly added under stirring. At this time, the mixture will heat up, so cooling and temperature control should be taken care of to prevent overheating. After the addition of the aromatic hydrocarbon is complete, stirring should continue for 1～2 hours at a certain temperature. If the reactivity of the aromatic hydrocarbon is high or moderate, 30 g（0.3 mol）of concentrated sulfuric acid can be used as the sulfonating agent, and the sulfonation process is the same as that for low-reactivity aromatic hydrocarbons. After sulfonation is complete, the reaction mixture is allowed to cool and then slowly poured into 100 mL of ice water under stirring. After the mixed solution is cooled, sodium carbonate is carefully added to make the solution neutral, followed by the addition of 30 g of sodium chloride to gradually precipitate the aromatic hydrocarbon sulfonic acid. The precipitate is filtered, washed with a small amount of water, dried, and placed in an oven for drying at 50～130℃.

【Note】

（1）The sulfonation temperature depends on the specific sulfonation target and the position of the sulfonic acid group entering the aromatic ring, and there are significant temperature differences. For example, at around 80℃, the main product of naphthalene sulfonation is α-naphthalenesulfonic acid, while at around 180℃, the main product is β-naphthalenesulfonic acid.

（2）The specific drying temperature depends on the dehydration temperature of the hydrated aromatic hydrocarbon sulfonic acid salt.

（3）Concentrated sulfuric acid has strong hygroscopicity and corrosiveness and can cause burns to the skin, so caution should be exercised during operations.

3.2.4 Experiment

Experiment Two *p*-Toluenesulfonyl chloride and *o*-Toluenesulfonyl chloride

【Experimental objectives】

To learn the principles and experimental methods of chlorosulfonation reactions, and to become familiar with the use of gas absorption apparatus, as well as recrystallization and vacuum distillation operations.

【Experimental principle】

CH_3 (toluene) $+ 2ClSO_3H \longrightarrow$ CH_3, SO_2Cl (o-isomer) $+$ CH_3, SO_2Cl (p-isomer)

p-Toluenesulfonyl chloride is a colorless oily liquid with a boiling point of 126℃/1.3 kPa（10 mmHg）. *o*-Toluenesulfonyl chloride forms crystalline flakes, with a melting point of 67-69℃ and a boiling point of 135℃/1.3 kPa（10 mmHg）.

【Chemicals】

Chlorosulfonic acid　　38 g（22 mL, 0.33 mol）

Toluene 10 g（12 mL, 0.11 mol）

【Experimental procedure】

Set up a stirrer, thermometer, dropping funnel, reflux condenser, and hydrogen chloride gas absorption apparatus on a 100 mL three-necked flask（refer to Figures 2-6 and 2-7）. Add 22 mL of chlorosulfonic acid to the three-necked flask and place the reaction vessel in an ice bath to cool it to 0°C. Note: Chlorosulfonic acid is highly corrosive and can react violently with water, even leading to explosions. It releases a large amount of hydrogen chloride gas when exposed to air. Therefore, the reaction apparatus and chemicals must be thoroughly dried, and caution must be exercised during operation. The measurements should be taken in a fume hood. Under stirring, slowly add 12 mL of anhydrous toluene to the reaction vessel from the dropping funnel. The dropping speed should be adjusted to maintain the reaction temperature below 5°C. If the temperature is too high, the generated *p*-toluenesulfonyl chloride will undergo hydrolysis. Controlling the dropping speed and ensuring sufficient stirring are crucial. After approximately 15 minutes, continue stirring at room temperature for 1 hour. Then, heat the mixture while stirring in a 40～50°C water bath until no more hydrogen chloride gas is evolved. After the reaction mixture has cooled to room temperature, slowly pour it into a beaker containing 80 mL of ice water, while stirring, in a fume hood. Rinse the reaction vessel with 20 mL of ice water, and combine the rinsing solution with the beaker. Then, remove the acid layer by tilting the beaker and separate the light yellow oily liquid, which is a mixture of *p*- and *o*-toluenesulfonyl chlorides. Wash the mixture twice with ice water, and then cool the oily mixture in a freezer at −10 to 20°C overnight（or it can be cooled using a calcium chloride-ice bath）. After freezing, the *p*-toluenesulfonyl chloride crystallizes out from the mixture. Perform suction filtration（preferably using a sand core funnel）, wash the filter cake with a small amount of cold water, and then perform another suction filtration to obtain crude p-toluenesulfonyl chloride. The filtrate mainly contains o-toluenesulfonyl chloride, which is extracted with chloroform. After washing the extraction solution with water, separate the organic phase and dry it with anhydrous magnesium sulfate. After evaporating the solvent, perform vacuum distillation and collect the fraction at 126°C/1.3 kPa（10 mmHg）. Alternatively, the product can be recrystallized with petroleum ether（30～60°C）.

【Notes】

（1）The boiling point of toluene is 110.8°C, and it forms an azeotrope with water. It boils at 84.1°C and contains 81.4% toluene. Toluene can be dried by azeotropic distillation, discarding the initial 20% of the distillate. If the water content is low, drying can also be achieved by adding anhydrous calcium chloride.

（2）If the reaction mixture is left in the refrigerator overnight, it may increase the proportion of p-toluenesulfonyl chloride.

（3）Sometimes, due to the low proportion of o-toluenesulfonyl chloride in the mixture, its crystallization is not easy to achieve. In such cases, crude p-toluenesulfonyl chloride can be obtained.

（4）Prior to vacuum distillation, sulfonic chlorides must be thoroughly dried, otherwise hydrolysis will occur under high-temperature conditions.

【Thought questions】

(1) What effect would it have on the reaction if the chemicals added before the chloro sulfonation reaction were not dried?

(2) Why is the reaction mixture poured into ice water after the chloro sulfonation reaction is complete?

(3) Based on what principle is the separation of the mixture of *p*-toluenesulfonyl and *o*-toluenesulfonyl chlorides in this experiment?

(4) If you were to synthesize methyl benzenesulfonic acid using toluene and chlorosulfonic acid as raw materials, how would you modify the steps of this experiment?

3.3 Nitration Reaction

The reaction in which hydrogen atoms in organic molecules are replaced by nitro groups (—NO_2) is called a nitration reaction. Nitration reactions on aromatic rings are an important class of electrophilic substitution reactions, it has a wide range of applications in fine organic synthesis. In the nitration reaction, the nitronium cation (NO_2^+) is the electrophilic reagent. The presence of a proton contributes to the generation of NO_2^+.

$$H^+ + HNO_3 \rightleftharpoons [H_2NO_3]^+$$

$$[H_2NO_3]^+ \rightleftharpoons H_2O + NO_2^+$$

Obviously, removing water molecules helps to increase the concentration of NO_2^+ and enhance nitration activity. In fact, adding concentrated sulfuric acid not only provides the necessary protons as shown in the above equation, but also absorbs water, thereby rapidly increasing the concentration of NO_2^+.

$$HNO_3 + 2H_2SO_4 \rightleftharpoons NO_2^+ + H_2O + 2HSO_4^-$$

However, for some easily nitrated aromatic rings, only diluted nitric acid is needed as the nitration agent, and the nitration conditions are relatively mild. These aromatic rings often contain first-class (activating) substituents (such as —OH, —OCH_3, —$NHCOCH_3$, —CH_3); aromatic rings with second-class (deactivating) substituents (such as —NO_2, —COOH, —SO_3H) require more severe nitration conditions and must use mixed acid composed of fuming nitric acid and concentrated sulfuric acid as the nitration agent.

m-Dinitrobenzene + fuming nitric acid (sulfuric acid) $\xrightarrow[5d]{110℃}$ 1,3,5-trinitrobenzene (NO_2, O_2N, NO_2)

1,2,4,5-Tetramethylbenzene (H_3C, CH_3, H_3C, CH_3) + fuming nitric acid (sulfuric acid) $\xrightarrow[15min]{50℃}$ dinitro-tetramethylbenzene (NO_2, H_3C, CH_3, H_3C, CH_3, NO_2)

3.3.1 General Procedure for Nitration of Aromatic Hydrocarbons

Different aromatic hydrocarbons with different reactivity require different compositions of nitration agents. Taking 0.1 mol of aromatic hydrocarbon compound as an example, for highly reactive aromatic hydrocarbons such as phenols and phenyl ethers, only 20 mL of 40% nitric acid（*d*=1.25, 0.2 mol）is needed. For moderately reactive aromatic hydrocarbons such as benzene, toluene, and naphthalene, a mixed acid composed of 8 mL of 68% concentrated nitric acid（*d*=1.41, 0.13 mol）and 10 mL of concentrated sulfuric acid is required. For low-reactivity aromatic hydrocarbons with deactivating groups, such as benzoic acid, nitrobenzene, a mixed acid composed of 10 mL of 100% concentrated nitric acid（*d*=1.51, 0.24 mol）and 14 mL of 100% concentrated sulfuric acid（*d*=1.83, 0.26 mol）is required.

Note: Nitric acid and sulfuric acid are highly corrosive and should be used with caution. When measuring nitric acid, it should be done in a fume hood. Multiple nitro compounds should never be distilled, even for distilling mononitro compounds, caution should be taken not to dry them to avoid explosions. Preparation of mixed acid: Under conditions of an ice bath, slowly add concentrated sulfuric acid to concentrated nitric acid while stirring, and cool the mixed acid to 10°C. Set up a stirrer, thermometer, dropping funnel. on a 250 mL three-necked flask, ensuring that the reaction system is open to the atmosphere（see Figure 2-6）. Add 0.1 mol of the aromatic hydrocarbon to the three-necked flask and slowly add the mixed acid from the dropping funnel while stirring. The reaction temperature is generally controlled at 20～50°C, and the reaction time is about 2～3 hours. For highly reactive aromatic hydrocarbons, the reaction temperature should be lower（0～10°C）. After the addition of the mixed acid is complete, stir at room temperature for 30 minutes. After the reaction is complete, slowly pour the reaction mixture into 200 mL of ice water while stirring. If the product is a solid at room temperature, filter and wash with water until neutral, then dry to obtain the crude nitration product, which can be further recrystallized using an appropriate solvent. If the product is a liquid, separate the organic layer using a separatory funnel, extract the acidic layer with ether, combine the extraction solution with the organic phase, wash successively with water, 10% sodium bicarbonate solution, and water until neutral, and then dry with calcium chloride before performing vacuum distillation.

3.3.2 Experiment

Experiment Three Para-Nitrophenol and Ortho-Nitrophenol

【Experimental objectives】

To learn the basic theory and methods of aromatic hydrocarbon nitration reaction, deepen the understanding of electrophilic substitution reactions in aromatic hydrocarbons, and master the technique of steam distillation.

【Experimental principle】

$$C_6H_5OH \xrightarrow{NaNO_3/H_2SO_4} o\text{-}O_2NC_6H_4OH + p\text{-}O_2NC_6H_4OH$$

Ortho-nitrophenol mp 45℃, has a distinctive aromatic odor.

Para-nitrophenol is a light yellow or colorless needle-like crystal, odorless, mp 112～113℃.

【Chemicals】

Phenol	4.7 g (0.05 mol)
Sodium nitrate	7 g (0.08 mol)
Concentrated sulfuric acid (d=1.83)	11 g (6 mL, 0.11 mol)
Concentrated hydrochloric acid	3 mL

【Experimental procedure】

In a 100 mL three-neck flask, set up a stirrer, thermometer, and dropping funnel (see Figure 2-6). Add 20 mL of water first, then slowly add 6 mL of concentrated sulfuric acid under stirring.

Note: Only pour the concentrated sulfuric acid slowly along the container wall into the water, do not reverse the order!

Remove the dropping funnel, while the acid is still warm, add 7 g of sodium nitrate from the side opening of the reaction flask to dissolve it in diluted sulfuric acid. Attach the dropping funnel and place the reaction flask in an ice-water bath to cool the mixture to 20℃.

Weigh 4.7 g of phenol, mix it with 1 mL of warm water, and cool it to room temperature.

Note: Phenol is corrosive. If it accidentally comes into contact with the skin, rinse immediately with soap and water, and then wipe with alcohol-soaked cotton.

After stirring, slowly add the phenol aqueous solution from the dropping funnel into the reaction flask, and maintain the reaction temperature around 20℃ using an ice-water bath.

Note: Good stirring in non-homogeneous reaction systems can significantly accelerate the reaction.

After adding phenol, continue stirring at room temperature for 1 hour. A black oily substance is formed, pour out the acid layer. Then add 20 mL of water to the oily substance and shake it, pour out the wash liquid first, then wash three times with water to remove the remaining acid.

Note: The nitrophenol product is toxic, be careful during the washing operation!

Perform steam distillation on the oily mixture until no yellow oil drops are distilled in the condenser. During the steam distillation process, yellow ortho-nitrophenol crystals will adhere to the inner wall of the condenser. You can occasionally close the cooling water tap to melt it and let it flow out with the hot steam.

Cool and filter the distillate, collect light yellow crystals to obtain the ortho-nitrophenol product. After air-drying, weigh, measure the melting point, and calculate the yield.

Note: Ortho-nitrophenol is volatile and should be stored in a sealed brown bottle.

Add water to the residue after steam distillation to a total volume of 50 mL, add 3 mL of concentrated hydrochloric acid and 0.5 g of activated carbon, boil for 15 minutes, filter through a preheated Buchner funnel, and the filtrate will cool and precipitate para-nitrophenol. After filtration and drying, weigh, measure the melting point, and calculate the yield.

If the measured melting point is too low, the product can be recrystallized using an ethanol-water mixed solvent: Add a small amount of ethanol to a round-bottom flask containing the para-nitrophenol, set up a reflux condenser, heat under reflux, and add ethanol until the product is completely dissolved in the boiling ethanol. Then, dropwise add hot water (around 60°C) until the ethanol solution becomes turbid. Add a few more drops of ethanol to make the turbid solution just clear. Allow it to cool to room temperature, filter to obtain the product, and measure the melting point after air-drying.

【Notes】

(1) The melting point of phenol is 41°C, and it is solid at room temperature. It can be melted by using a warm water bath when measuring. Adding a small amount of water to phenol can lower its melting point, making it liquid at room temperature, which is beneficial for dripping and reaction.

(2) The reaction temperature has a significant impact on the nitration of phenol. When the temperature is too high, mononitrophenol may undergo further nitration, or the yield of mononitrophenol may decrease due to oxidation reactions. When the temperature is too low, the reaction rate will be slowed down.

(3) During the steam distillation of nitrophenols in the residual mixed acid, further nitration or oxidation may occur due to prolonged high-temperature heating. Therefore, it is necessary to wash away the residual acid in the crude product.

【Thought questions】

(1) What possible side reactions may occur during the nitration of phenol?

(2) Why can ortho-nitrophenol and para-nitrophenol be separated by steam distillation? Can the same method be used to separate ortho-nitrotoluene and para-nitrotoluene?

3.4 Friedel-Crafts Reaction

3.4.1 Friedel-Crafts Reaction Principle

In 1877, French chemist Charles Friedel (C. Friedel) and American chemist James Crafts (J.M. Crafts) discovered during their collaborative research that when anhydrous aluminum trichloride was added to a stable mixture of chloromethane and benzene, a violent reaction occurred, releasing a large amount of hydrogen chloride gas. They isolated toluene from the reaction mixture. Subsequent studies showed that this type of reaction has general significance. It is not only applicable to alkylating reactions of aromatic hydrocarbons but also to acylation reactions of aromatic hydrocarbons. This later became known as the Friedel-Crafts alkylation and acylation reactions, also known as the Friedel-Crafts reaction.

$$C_6H_6 + RCl \xrightarrow{AlCl_3} C_6H_5-R$$

$$C_6H_6 + RCOCl \xrightarrow{AlCl_3} C_6H_5-\overset{O}{\overset{\|}{C}}-R$$

Alkylation and acylation reactions of aromatic hydrocarbons are widely used for synthesizing alkylaromatic hydrocarbons and aromatic ketones.

In alkylating reactions, in addition to alkyl halides, alkenes and alcohols can also be used as alkylating agents. Due to the activating effect of alkyl groups on aromatic hydrocarbons, in the Friedel-Crafts alkylation reaction, alkylaromatic hydrocarbons are more prone to undergo further alkylation than the starting aromatic hydrocarbons, resulting in the formation of polyalkylaromatic hydrocarbons. Therefore, to obtain monosubstituted aromatic hydrocarbons, an excess of aromatic hydrocarbons must be added to control the formation of polyalkylaromatic hydrocarbons. Here, the aromatic hydrocarbon serves as both the reactant and diluent. If the aromatic hydrocarbon is a solid, an inert medium such as carbon disulfide, petroleum ether, or nitrobenzene can be added separately. In addition, due to the formation and attack of alkyl carbocations in alkylating reactions, rearrangement products can be formed. This limits the application of alkylating reactions in the synthesis of linear alkylaromatic hydrocarbons with more than two carbon atoms.

In contrast to alkylation reactions, acylation reactions of aromatic hydrocarbons stop at the monosubstituted stage due to the deactivating effect of acyl groups on the aromatic ring. This is advantageous for selectively preparing monosubstituted aromatic hydrocarbons. Furthermore, since no rearrangement occurs in acylation reactions, they have special value in the synthesis of linear alkylaromatic hydrocarbons or alkylaromatic hydrocarbons with other branched structures. For example, normal propylbenzene can be obtained through acylation and reduction reactions, which not only stops at the monosubstituted stage but also retains the original carbon chain structure.

$$C_6H_6 + CH_3CH_2COCl \xrightarrow{AlCl_3} C_6H_5COCH_2CH_3 \xrightarrow{[H]} C_6H_5CH_2CH_2CH_3$$

Common acylating agents include acyl chlorides and acid anhydrides.

In the Friedel-Crafts reaction using alkyl halides as alkylating agents, common catalysts include anhydrous aluminum trichloride, zinc chloride, and boron trifluoride, which are Lewis acids. Among them, anhydrous aluminum trichloride has the best catalytic efficiency（in alkylating reactions using alkenes or alcohols as alkylating agents, proton acids such as hydrogen fluoride, sulfuric acid, and phosphoric acid are generally used as catalysts）. In alkylating reactions, the amount of anhydrous aluminum trichloride added is only the amount of catalyst, but in acylation reactions, it is different. Since anhydrous aluminum trichloride can form stable complexes with carbonyl compounds, the amount of anhydrous aluminum trichloride used as a catalyst alone is not enough. In the case of acyl chlorides as acylating agents, considering that both the acyl chloride and the acylation product (aromatic ketone) can form complexes with aluminum trichloride, 1 mole of acyl chloride requires more than 1 mole of anhydrous aluminum trichloride to be added, usually with an excess of 10%. If acid anhydrides are used as acylating agents, even more anhydrous aluminum trichloride is needed. This is because in the Friedel-Crafts reaction, acetic acid is generated from acid anhydrides, and both acetic acid and the acylation product aromatic ketone consume an equimolar amount of aluminum trichloride to form complexes. Therefore, 1 mole of acid anhydride requires at least 2 moles of aluminum trichloride, and in practical preparations, an additional 10% excess is usually used. In acylation reactions, excess aromatic hydrocarbons or solvents such as carbon disulfide, nitrobenzene, and petroleum ether are often used.

3.4.2 Alkylation Experimental Procedure

In terms of the monoalkylation reaction of benzene, benzene is greatly in excess during the reaction, so the amount of haloalkane is used as a reference.

For a feed amount of 0.1 mol haloalkane, a 100 mL three-necked flask can be used as the reaction vessel. It is equipped with a mechanical stirrer, dropping funnel, thermometer, and reflux condenser. A calcium chloride drying tube is attached to the upper end of the condenser, which is connected to a gas absorption device (see Figure 2-6, Figure 2-7).

First, add 1.3 g (0.01 mol) anhydrous aluminum trichloride and 30 mL dry benzene into the three-necked flask, which has been ground into powder. Under stirring, slowly add a solution prepared by mixing 0.1 mol haloalkane and 15 mL dry benzene from the dropping funnel. The reaction generates heat by itself, so it is important to control the rate of addition. The dripping speed should be adjusted to maintain a reaction temperature of 20~25℃, if necessary, an ice-water bath can be used for cooling. After the addition is complete, continue stirring at room temperature until the reaction tends to stabilize. Then, raise the bath temperature to 60~70℃ and heat and stir until no more hydrogen chloride gas escapes from the reaction mixture. After the reaction mixture has cooled, slowly pour it into 50 mL of ice water containing crushed ice in a fume hood. If there is still precipitate, add a small amount of hydrochloric acid to dissolve it, and stir thoroughly with a glass rod. Then, use a separatory funnel to separate the aqueous phase, and wash the organic phase successively with water, 5% sodium carbonate solution, and water until the organic phase becomes neutral. After drying with anhydrous calcium chloride, evaporate the solvent and collect the product through distillation (or recrystallization).

3.4.3 Acylation Experimental Procedure

For a feed amount of 0.1 mol aromatic hydrocarbon, a 250 mL three-necked flask can be used as the reaction vessel. It is equipped with a mechanical stirrer, dropping funnel, and a reflux condenser. A calcium chloride drying tube is attached to the upper end of the condenser, which is connected to a gas absorption device (see Figure 2-6, Figure 2-7 for reference). First, add

16 g（0.12 mol）anhydrous aluminum trichloride, 100 mL dried petroleum ether（60～90℃）, and 0.1 mol aromatic hydrocarbon into the three-necked flask. Under stirring, slowly add 0.1 mol acyl chloride, and control the reaction temperature to be around 40℃. After the addition is complete, heat and stir in a water bath at 50～60℃ for 1～2 hours until no more hydrogen chloride gas escapes from the reaction mixture. After the reaction mixture has cooled, slowly pour it into 100 mL of ice water in a fume hood, and aluminum hydroxide precipitates. Add 5～10 mL concentrated hydrochloric acid while stirring to dissolve the precipitate. Then, separate the organic phase using a separatory funnel and extract the water phase twice with petroleum ether or dichloroethane. Combine the organic phases and wash the combined organic phase successively with water, 5% sodium hydroxide solution, and water until it becomes neutral. After drying with anhydrous magnesium sulfate, evaporate the solvent and collect the product through vacuum distillation. If acetic anhydride is used as the acylating reagent, the acylation experimental steps are similar to acyl chloride-based acylation, except that the amount of anhydrous aluminum trichloride should be increased. That is, 32 g（0.24 mol）anhydrous aluminum trichloride should be added instead of 0.1 mol acyl chloride. Note: If the aromatic hydrocarbon is in a liquid state at room temperature, it can be replaced with an equal volume of aromatic hydrocarbon instead of other solvents.

3.4.4 Experiment One

Experiment Four Benzoin (acylation method) (Sweet-flavored spice)

【Experimental objectives】

To learn the theory and experimental methods of the Friedel-Crafts acylation reaction, and master the techniques of extraction, distillation, and vacuum distillation.

【Experimental principle】

$$C_6H_5-\overset{O}{\overset{\|}{C}}-Cl + C_6H_6 \xrightarrow{AlCl_3} C_6H_5-\overset{O}{\overset{\|}{C}}-C_6H_5$$

Dibenzoin is a colorless crystals, with a melting point of 47～48℃ and a boiling point of 305.4℃.

【Chemicals】

Anhydrous aluminum trichloride	7.5 g（0.056 mol）
Anhydrous benzene	27 g（30 mL, 0.34 mol）
Benzoyl chloride	7.3 g（6 mL, 0.05 mol）
5% sodium hydroxide solution	20 mL
Concentrated hydrochloric acid	2～3 mL

【Experimental procedure】

Set up a stirring device, constant-pressure dropping funnel, thermometer, and reflux condenser on a 100 mL three-necked flask. Install a calcium chloride drying tube and a gas absorption device filled with alkaline solution on the upper end of the condenser (see Figure 2-6, Figure 2-7).

Note: All instruments and chemicals used in the experiment should be dry.

Weigh 7.5 g of anhydrous aluminum trichloride in a weighing boat inside a fume hood. Grind it finely and quickly transfer it into the three-necked flask, and add 30 mL of benzene. While stirring at room temperature, slowly add 6 mL of benzoyl chloride to the flask using the constant-pressure dropping funnel. Control the reaction temperature to around 40℃ by adjusting the dropping speed.

Note: Benzoyl chloride is lachrymatory and can cause strong irritation to the skin, eyes, and respiratory tract. Therefore, it should be handled inside a fume hood.

The mixture in the flask will start to react vigorously, accompanied by the generation of hydrogen chloride gas. The reaction mixture gradually turns brown. After approximately 10 minutes, finish the dropwise addition and heat the flask while stirring in a 60℃ water bath until no more hydrogen chloride gas escapes from the reaction mixture. This process usually takes about 1.5 hours.

After the three-necked flask has cooled down, slowly pour the reaction mixture into a beaker containing 50 mL of ice-cold water inside a fume hood. A precipitate will form. Stir the mixture, and slowly add 2～3 mL of concentrated hydrochloric acid using a dropper until the precipitate is completely dissolved. Separate the organic phase using a separatory funnel, and extract the aqueous phase twice with benzene (2×15 mL). Combine the organic phases, wash them sequentially with 20 mL of water and 20 mL of 5% sodium hydroxide solution, and then wash them with water 2～3 times (each time with 20 mL) until the organic phase is neutral. Dry the organic phase with anhydrous magnesium sulfate and evaporate the solvent to obtain the crude product. Perform vacuum distillation to collect the fraction at 187～190℃/2.0 kPa (15 mmHg), and let it solidify after cooling to obtain the pure product.

If vacuum distillation is not performed, the crude product can be recrystallized using petroleum ether (60～90℃). Weigh and measure the melting point of the dried product to calculate the yield.

【Notes】

(1) Redistill benzene and discard the first 10% fraction to meet the requirements.

(2) Anhydrous aluminum trichloride is extremely hygroscopic. It will produce irritating hydrogen chloride gas when in contact with moist air. Weighing, grinding, and feeding operations should be performed quickly.

(3) The crude product often appears viscous due to the incomplete removal of the solvent or the presence of different crystal forms, resulting in a decrease in melting point.

(4) Dibenzoin has multiple crystal forms, and their melting points are different: α-form is at 49℃, β-form is at 26℃, γ-form is at 45～48℃, and δ-form is at 51℃. Among them, the α-form is more stable.

【Thought questions】

(1) Why is an excess of anhydrous aluminum trichloride used as a catalyst in the Friedel-Crafts acylation reaction using acyl chlorides as acylating agents?

(2) Why is acid treatment performed after the acylation reaction?

(3) Is it easy to produce polyacylated aromatic hydrocarbons in the acylation reaction?

3.4.5 Experiment Two

Experiment Five Benzophenone (alkylation method)

【Experimental objectives】

To study the theory and experimental methods of Friedel-Crafts alkylation reaction, and to master the techniques of steam distillation, extraction, and vacuum distillation.

【Experimental principle】

$$C_6H_6 \xrightarrow[AlCl_3]{CCl_4} C_6H_5-CCl_2-C_6H_5 \xrightarrow{H_2O} C_6H_5-CO-C_6H_5$$

【Chemicals】

Anhydrous benzene	7.8 g (9 mL, 0.1 mol)
Carbon tetrachloride	34 g (22 mL, 0.22 mol)
Anhydrous aluminum trichloride	6.7 g (0.05 mol)
Anhydrous magnesium sulfate	1～2 g Benzene: 20 mL

【Experimental procedure】

Set up a mechanical stirrer, dropping the funnel, thermometer, and reflux condenser on a 250 mL three-necked flask. Attach a calcium chloride drying tube and gas absorption unit on the upper end of the condenser (see Figure 2-6, Figure 2-7).

Note: The instruments and chemicals used must be dried beforehand.

Add 6.7 g of finely ground anhydrous aluminum trichloride and 15 mL of carbon tetrachloride quickly to the three-necked flask.

Note: Carbon tetrachloride is toxic! Avoid inhalation of its vapors. Anhydrous aluminum trichloride is hygroscopic, so the operation should be quick.

Place the three-necked flask in an ice-water bath until the temperature inside the flask drops to about 12°C. Under stirring, slowly add 4 mL of a solution made from 9 mL of anhydrous benzene and 7 mL of carbon tetrachloride. Chlorine gas is generated when the reaction starts, and the temperature of the reaction mixture gradually increases. Control the reaction temperature to 12°C using an ice-water bath.

When the reaction becomes milder, add the remaining benzene solution dropwise into the reaction flask, maintaining the reaction temperature between 5～10°C. After the addition is complete (takes about 15 minutes), continue stirring for 1 hour while keeping the reaction temperature around 10°C. After the reaction is complete, change the reaction setup to a steam distillation setup (see Figure 2-11). Slowly add 40 mL of water dropwise into the three-necked flask through the dropping funnel. The reaction mixture gradually heats up. Control the rate of water addition to maintain the stable boiling and direct distillation of excess carbon tetrachloride. After adding water, heat the three-necked flask over low heat on an asbestos net to 0.5 hours to remove the residual carbon tetrachloride (if too much water evaporates from the flask, add an appropriate amount of water through the dropping funnel). This step also promotes complete hydrolysis of dichlorodiphenylmethane. Then transfer the mixture in the three-necked flask to a separatory funnel, separate the organic phase, and extract the aqueous phase with 20 mL of benzene. Combine the extraction liquid with the organic phase, wash the combined liquid with water until neutral, dry it with magnesium sulfate, distill off the solvent under atmospheric pressure, and then perform vacuum distillation to collect the fraction at 187～190°C/2.0 kPa (15 mmHg). The product solidifies upon cooling and has a melting point of 47～48°C.

【Notes】

(1) Anhydrous carbon tetrachloride and anhydrous benzene that meet the requirements of the Friedel-Crafts reaction can be obtained by simple distillation, discarding the first 10% of the distillate.

(2) When the reaction temperature is lower than 5°C, the reaction is too slow. When the temperature is higher than 10°C, tar-like resin products are easily formed.

(3) Since heating is required for the hydrolysis of the intermediate dichlorodiphenylmethane, there is no need to cool during the decomposition of the aluminum trichloride complex, allowing the intermediate to undergo preliminary hydrolysis.

【Thought questions】

(1) In the Friedel-Crafts reaction, what are the differences in the use of anhydrous aluminum trichloride as a catalyst between alkylation and acylation reactions? Why?

(2) What possible side reactions may occur in this experiment, and what measures are taken to reduce them?

(3) Theoretically, for the preparation of 1 mol of benzophenone in this experiment, 2 mol of benzene and 1 mol of carbon tetrachloride are needed. What is the actual feed ratio? Please explain briefly.

3.5 Oxidation Reaction

3.5.1 Principles of Oxidation Reaction

In chemical reactions, any reaction that loses electrons is called an oxidation reaction. The oxidation reaction of organic compounds is characterized by a decrease in the number of hydrogen atoms or an increase in the number of oxygen atoms in the molecule. In organic synthesis, oxidation reaction is an important class of unit reaction. Through oxidation reactions, many oxygen-containing compounds such as alcohols, aldehydes, ketones, acids, phenols, quinones, and epoxides can be obtained, and they have a wide range of applications. In industry, inexpensive air or pure oxygen is often used as an oxidant, but due to its weak oxidation ability, the oxidation reaction usually occurs under high temperature and high pressure conditions. The oxidants commonly used in the laboratory include potassium permanganate, sodium dichromate, nitric acid, etc. These oxidants have strong oxidation ability and can oxidize various functional groups, belonging to universal oxidants.

When potassium permanganate is used as the oxidant, the reaction results are not the same in different media. When potassium permanganate is oxidized in neutral or alkaline media, the valence state of manganese atoms decreases from +7 to +4, generating manganese dioxide, which is insoluble in water and precipitates. During this process, on average 1.5 mol of atomic oxygen is released for every 1 mol of permanganate ion:

$$2KMnO_4 + H_2O \longrightarrow 2MnO_2 \downarrow + 2OH + 3[O]$$

In acidic media, the valence state of manganese atoms drops from +7 to +2, forming divalent manganese salts. On average, 1 mol of permanganate ion releases 2.5 mol of atomic oxygen:

$$2KMnO_4 + 3H_2SO_4 \longrightarrow 2MnSO_4 \downarrow + K_2SO_4 + 3H_2O + 5[O]$$

For example, to oxidize 1 mol of ethylbenzene to benzoic acid under alkaline conditions, 4 mol of potassium permanganate is needed:

$$2KMnO_4 + 3H_2SO_4 \longrightarrow 2MnSO_4 \downarrow + K_2SO_4 + 3H_2O + 5[O]$$

Relatively speaking, the oxidation reaction of potassium permanganate in neutral media is somewhat milder and is suitable for preparing catechols from olefins.

$$CH_3(CH_2)_7CH{=}CH(CH_2)_7COOH \xrightarrow{KMnO_4,\ H_2O} CH_3(CH_2)CH(OH)CH(OH)(CH_2)_7COOH$$

In alkaline media, potassium permanganate can oxidize primary alcohols or aldehydes to their corresponding acids, and can also be used to oxidize side chains on aromatic hydrocarbons.

$$CH_3CH(CH_3)CH_2OH \xrightarrow{KMnO_4,\ OH^-} CH_3CH(CH_3)COOH$$

$$p\text{-}CH_3C_6H_4COCH_3 \xrightarrow{KMnO_4,\ OH^-} p\text{-}HOOCC_6H_4COOH$$

The oxidation reaction of potassium permanganate in acidic media is usually carried out in a sulfuric acid solution of more than 25%. In acidic media, the oxidation of alkylated aromatic hydrocarbons with potassium permanganate is often accompanied by decarboxylation reactions, so the acidic media method is rarely used. To determine the ratio of the reactant to $KMnO_4$ for oxidation, taking the oxidation of toluene to benzoic acid under alkaline conditions as an example, 3 mol of atomic oxygen is required to oxidize 1 mol of toluene, which means that 2 mol of $KMnO_4$ is needed for 1 mol of toluene:

$$C_6H_5CH_3 + 3[O] \longrightarrow C_6H_5COOH + H_2O$$

Potassium permanganate is generally insoluble in organic solvents except for a few solvents such as acetone, acetic acid, and tert-butanol. Its oxidation reactions are mostly carried out in aqueous solutions. When using potassium permanganate in aqueous solutions to oxidize organic compounds, the reaction occurs at the interface between the water phase and the organic phase, and the reaction rate is relatively slow, and the yield is also relatively low. If some phase transfer catalysts are added, the effect will be remarkable. As an oxidant, 1 mol of nitric acid can release 1.5 mol of atomic oxygen:

$$2HNO_3 \longrightarrow 2NO_2 + H_2O + 3[O]$$

Because nitric acid has strong oxidation properties, it is usually used in the preparation of carboxylic acids. For example, adipic acid can be obtained by oxidizing cyclohexanol with nitric acid to produce adipic acid.

3.5.2 General Procedure of Potassium Permanganate Oxidation Experiment

The potassium permanganate oxidation reaction is mostly carried out under alkaline conditions. Taking the oxidation of alkylated aromatic hydrocarbons to aromatic carboxylic acids as an example, based on a reactant input of 0.1 mol of alkylated aromatic hydrocarbons, 0.1 mol of sodium carbonate and 600 mL of water are added to a 1000 mL three-necked flask. The mixture is stirred, and 0.3 mol of potassium permanganate (the amount of potassium permanganate is measured based on the relationship between the formation of a carboxyl group from a methyl group on the aromatic hydrocarbon) is gradually added to the reaction flask during heating and refluxing. After each addition of potassium permanganate, the reaction solution should be allowed to fade to colorless before adding another batch. After all the reagents are added, if the purple color does not disappear after a while, an appropriate amount of potassium permanganate (about 10% of the total input) should be added, and then stirring and refluxing for 2~4 hours.

The mixture is filtered while hot, and the manganese dioxide precipitate is washed three times with boiling water. The filtrate is combined, acidified with 50% sulfuric acid, and the aromatic acid precipitates are obtained by filtration. The precipitate is then washed with a small amount of cold water and recrystallized with water.

3.5.3 Experiment

Experiment Six Nicotinic Acid (Anti-pellagra drug)

【Experimental objectives】

To study the oxidation principles and experimental methods of aliphatic aromatic hydrocarbons using the potassium permanganate oxidation method, and to learn the techniques for separation and purification of water-soluble organic compounds.

【Experimental principle】

$$\text{3-methylpyridine (}C_5H_4N\text{–}CH_3\text{)} + 2KMnO_4 \longrightarrow \text{nicotinic acid (}C_5H_4N\text{–}COOH\text{)} + 2MnO_2 + KOH + H_2O$$

Nicotinic acid is a colorless needle-like crystal with a melting point of 236-239℃.

【Chemicals】

3-Methylpyridine	3.0 g（3.1 mL, 0.032 mol）
Potassium permanganate	12 g（0.075 mol）
Concentrated hydrochloric acid	12～13 mL

【Experimental procedure】

Set up a stirrer, powder funnel, and thermometer on a 250 mL three-necked flask. Add 3 g of 3-methylpyridine and 100 mL of water to the flask. Place the flask in a water bath and heat it to 70℃. Stir the mixture and divide 12 g of potassium permanganate into 10 portions for gradual addition. After each portion of potassium permanganate is added, wait for the purple color of the reaction solution to fade before adding the next portion. Initially, maintain the reaction temperature at 70℃. After adding 6 g of potassium permanganate, increase the reaction temperature to 85～90℃, and then add the remaining 6 g of potassium permanganate in batches. After all the substances are added, heat the flask in a boiling water bath while stirring. Filter the solution while it is still hot after the purple color of potassium permanganate fades. Wash the manganese dioxide filter cake with hot water 3～4 times（10 mL each time）. Combine the filtrate in a beaker and heat and concentrate the filtrate to about 100 mL. Then, add concentrated hydrochloric acid dropwise to the concentrated solution（about 4 mL）until the pH of the solution reaches 3.4（isoelectric point of nicotinic acid）. Note: Test with precision pH paper.

Allow the solution to cool slowly（or refrigerate overnight）to allow nicotinic acid crystals to gradually precipitate. Filter and collect the solid product, wash it with a small amount of cold water, and dry the crude product under 90～100℃ conditions. Evaporate the filtrate to a volume of 60 mL, then slowly cool it to 5℃ to obtain the second batch of product.

The crude product can be recrystallized with water.

【Notes】

Slow cooling during crystallization helps reduce the impurity content of potassium chloride in the product.

【Thought questions】

（1）The solubility of nicotinic acid in water（g/100 mL）is as follows: 0.10（0℃）, 0.26（40℃）, 0.82（80℃）, 1.27（100℃）Propose an experimental plan for recrystallizing 2 g of crude nicotinic acid.

（2）Why is the pH adjusted to the isoelectric point of nicotinic acid during post-processing of the product?

（3）Why is it emphasized to slowly cool the second concentrated solution during the post-processing of the reaction mixture? What would happen if the cooling rate is too fast?

（4）If the nicotinic acid product still contains a small amount of potassium chloride impurity, how can it be removed? Propose a separation and purification plan.

3.6 Reduction Reactions

In organic synthesis chemistry, a type of reaction that involves the addition of hydrogen atoms or the removal of oxygen atoms from organic molecules is called a reduction reaction. Reduction reactions play an important role in fine organic synthesis. Common methods of reduction include metal reduction using proton donors such as acids and alcohols, catalytic hydrogenation, and metal hydride reduction. Among them, the use of metals as reducing agents in combination with proton donors is widely applied in laboratory settings. Metals such as lithium, sodium, potassium, magnesium, zinc, tin, iron, etc., which have an electrode potential higher than that of hydrogen, can be used as reducing agents. Common proton donors include acids, alcohols, water, ammonia, etc.

If the reaction between the metal and proton donor is too vigorous, the reduction effect may not be satisfactory because the protons will escape in the form of molecular hydrogen. For example, metallic sodium cannot be used as a reducing agent with hydrochloric acid, but it can be used as a reducing agent with alcohols.

$$H_3C{-}C(=O){-}C_5H_{11} \xrightarrow{Na+CH_3CH_2OH} CH_3CH(OH)C_5H_{11}$$

There are numerous combinations of metals with an electrode potential higher than that of hydrogen and different proton donors, and their reduction performance is closely related to reaction conditions and the structure of the substance being reduced. Therefore, it is not convenient to summarize the entire picture with a single experimental method here.

3.6.1 General Procedure for Carbonyl Reduction (Huang Minglong Reduction)

In a 500 mL three-necked flask, equipped with a thermometer, stirrer, and condenser (see Figure 2-6), add 0.1 mol of carbonyl compound, 13 g (0.23 mol) of potassium hydroxide, and 180 mL of diethylene glycol to the reaction flask. Start stirring and heat to 130°C, maintaining this temperature for 2 hours. Then, change the reaction apparatus to a distillation apparatus and distill off the excess hydrazine and water until the temperature of the reaction mixture rises to 195～210°C. Switch the apparatus to a reflux setup and maintain reflux until no nitrogen gas is released (approximately 4 hours). After cooling, dilute the reaction mixture with an equal volume of water, and then extract with diethyl ether three times. Wash the diethyl ether extracts successively with diluted hydrochloric acid, water, 5% sodium carbonate solution, and water. Dry the extracts with anhydrous magnesium sulfate and remove the solvent under vacuum. Finally, distill or recrystallize the residue.

3.6.2 General Procedure for Nitroaromatic Iron Filings Reduction

Generally, the reduction of 1 mol of nitro compound requires 3-4 mol of iron filings, which greatly exceeds the theoretical value. In the reaction process using iron filings as the reducing agent, the presence of an electrolyte can improve the conductivity of the solution, accelerate the corrosion process of iron, and speed up the reduction rate. Studies have shown that in the reduction reaction of nitrobenzene with iron filings, different electrolytes affect the reduction rate in the following order of activity:

$$NH_4Cl>FeCl_2>(NH_4)_2SO_4>BaCl_2>CaCl_2>NaCl>Na_2SO_4>KBr>NaAC$$

Water acts as both the medium and the hydrogen source in the reduction of nitroaromatics by iron filings. The ratio of water to nitroaromatic compound is 50～100: 1. For less active nitroaromatics, methanol, ethanol, or other solvents can be added to the reaction mixture to facilitate the reaction.

Set up a reflux condenser on a round-bottom flask and add 0.3 mol of iron powder, 2 g of ammonium chloride, and 50 mL of water subsequently. Stir and heat gently until boiling for 15 minutes. Then add 0.1 mol of nitroaromatic compound. Stir and reflux for 1.5 hours. Cool to room temperature for further treatment. Depending on the properties of the amine products, corresponding separation and purification methods can be employed as follows:

(1) For aromatic amines that are insoluble in water and have a certain vapor pressure, distillation using steam can be used for separation. Examples include aniline, ortho-toluidine, meta-toluidine, ortho-chloroaniline, meta-chloroaniline, etc.

(2) For water-soluble aromatic amines that can be distilled, separation can be achieved by filtering out the iron sludge, simple distillation to remove water, and finally carrying out vacuum distillation. Examples include *o*-phenylenediamine, *p*-phenylenediamine, 2, 4-diaminotoluene, etc.

(3) For hot-water-soluble aromatic amines, separation can be achieved by hot filtration followed by cooling and crystallization. Examples include ortho-phenylenediamine, ortho-aminophenol, para-aminophenol, etc.

(4) For aromatic amines that are insoluble in water and have very low vapor pressure, extraction with a solvent can be used. Examples include alpha-naphthylamine.

【Notes】

Ammonium chloride hydrolyzes to form hydrochloric acid. Treating the iron filings with dilute hydrochloric acid can enhance the reaction activity.

3.7 Williamson Reaction

3.7.1 Principle of Williamson Reaction Experiment

There are various methods for the preparation of ethers, such as alcohol dehydration, sulfuric acid dialkyl ester and phenolate reaction, or Williamson reaction. The alcohol dehydration method is commonly used to prepare simple ethers (also known as symmetrical ethers) such as methyl ether and ethyl ether. If two different alcohols are dehydrated to prepare mixed ethers (also known as unsymmetrical ethers), a mixture of several ethers will be generated, making separation difficult. Except for the preparation of aryl alkyl ethers, the dehydration method is generally not used to prepare mixed ethers. Due to the high toxicity of alkylating reagents such as dimethyl sulfate and diethyl sulfate, the Williamson reaction is commonly used for the synthesis of mixed ethers. It involves the nucleophilic substitution reaction of alkyl halides and sodium alkoxides to produce ethers.

$$RX + R'OM \longrightarrow ROR' + MX$$

In the Williamson reaction, the reactivity of alkyl halides follows the order: RI > RBr > RCl. Among them, alkyl iodides have the highest reactivity. When reacting RBr or RCl in acetone or alcohol solution, the addition of 10% (relative to the amount of alkyl halide) potassium iodide or sodium iodide can often accelerate the reaction. In addition, the size of the alkyl group in the alkyl halide molecule also has a significant influence on the reaction pathway. Generally, as the number of branches in the alkyl group increases, the tendency to form alkenes also increases. For example, when reacting tert-alkyl halides with sodium alkoxides, the main product is usually an alkene rather than an ether.

$$(CH_3)_3C\text{-}Br + NaOCH_3 \longrightarrow (H_3C)_2C{=}CH_2 + CH_3OH + NaX$$

Therefore, in the synthesis of mixed ethers with secondary or tertiary alkyl groups, the alkyl groups are often introduced in the form of alkoxide rather than alkyl halides. Taking the preparation of tert-butyl ether as an example.

$$(CH_3)_3C\text{-}ONa + ICH_3 \longrightarrow (CH_3)_3C\text{-}OCH_3 + NaI$$

The Williamson reaction can be used to prepare higher molecular weight alkyl ethers and mixed ethers, as well as alkyl aryl ethers. In the preparation of mixed ethers containing aromatic hydrocarbons, due to the low reactivity of halogens in the aromatic halide molecule, it is difficult to undergo substitution reactions. Meanwhile, phenols are more acidic than alcohols and easily form salts. Therefore, the route of reacting phenolate with alkyl halides should be adopted for synthesis.

$$C_6H_5ONa + CH_3I \longrightarrow C_6H_5OCH_3 + NaI$$

Due to the strong acidity of phenols, in the preparation of alkyl aryl mixed ethers, phenolate salts can be prepared using phenol and strong base, followed by the Williamson reaction.

$$C_6H_5OH + NaOH \longrightarrow C_6H_5ONa + H_2O$$

3.7.2 General Procedure of Williamson Reaction Experiment

Set up a dropping funnel and a reflux condenser with a calcium chloride drying tube in a three-necked flask. Add 2.0 g (0.13 mol) of sliced metallic sodium into the flask, then add 1 mol of dried alcohol dropwise. The rate of alcohol addition should maintain smooth boiling of the solution.

Note: Do not dispose of waste metallic sodium scraps in the sink to prevent accidents. The sodium scraps can be soaked in isopropanol for disposal. The reaction involving metallic sodium should never be heated with a water bath.

After complete dissolution of metallic sodium, add 0.1 mol of alkyl halide (or 0.1 mol of sulfuric acid dialkyl ester) dropwise into the three-necked flask through the dropping funnel. Heat the flask in an oil bath under reflux for 2 hours to promote the reaction.

Note: If the reactivity of the used alkyl halide is too low, such as bromoalkane, add about 1 g of anhydrous potassium iodide simultaneously while adding the alkyl halide.

For the preparation of aryl ethers, such as phenyl ether, follow the procedure below: Add 4.4 g (0.11 mol) of sodium hydroxide and 30 mL of water into a round-bottom flask. After complete dissolution of sodium hydroxide, add 9.4 g (0.1 mol) of phenol and stir quickly to dissolve it. Then add 0.1 mol of alkyl halide and heat under reflux for 2 hours. After cooling the reaction mixture, pour it into 50 mL of water, extract it with ether three times, combine the organic phase, wash it with water, and dry it with calcium chloride. Filter, evaporate the ether, and then distill or carry out vacuum distillation or recrystallization on the residue.

If it is necessary to recover the unreacted phenol, follow the procedure below: Acidify the alkaline aqueous solution obtained after extraction with diluted hydrochloric acid, extract it with ether, wash and dry it, then evaporate the ether to recover the unreacted phenol.

3.7.3 Experiment

Experiment Seven Methyl tert-butyl ether (MTBE) - An Anti-Knock Agent in Lead-Free Gasoline

【Experimental objectives】

To learn the principle and experimental method of Williamson ether synthesis.

【Experimental principle】

$$(CH_3)_3C-OH + HOCH_3 \xrightarrow{15\%\ H_2SO_4} (CH_3)_3C-OCH_3$$

Methyl tert-butyl ether is a colorless and transparent liquid with a boiling point of 55～56℃, refractive index of 1.3690, and a density of 0.740.

【Chemicals】

Tert-butanol	14.8 g（19 mL, 0.2 mol）
Methanol	12.8 g（16 mL, 0.4 mol）
15% sulfuric acid	70 mL
Anhydrous sodium carbonate	3～5 g

【Experimental procedure】

Set up a fractionating column on a 250 mL round-bottom flask, with a thermometer at the top of the column. Attach a straight condenser, a reflux condenser, and a collection bottle to the sidearm of the fractionating column. Connect the sidearm of the reflux condenser to a water bath. Place the collection bottle in an ice bath.

Add 70 mL of 5% sulfuric acid, 16 mL of methanol, and 19 mL of tert-butanol to the round-bottom flask. Shake well to mix the reagents. Add a few pieces of boiling stones and heat gently. Collect the fraction between 49～53℃.

Transfer the collected liquid to a separatory funnel and wash it successively with water, 10% Na_2SO_3 solution, and water to remove any alcohols and peroxides in the ether layer. The ether layer should appear clear and transparent after the alcohol is removed. Then dry the ether layer with anhydrous sodium carbonate and collect the fraction between 53～56℃. Weigh the product, measure its refractive index, and calculate the yield.

【Notes】

Tert-butanol has a melting point of 25.5℃ and a boiling point of 82.5℃. It is a liquid in the presence of a small amount of water. If it is difficult to add at room temperature due to low temperature, a small amount of water can be added to liquefy it before adding it to the reaction mixture.

【Thought questions】

（1）Generally, the Williamson synthesis method is suitable for preparing mixed ethers. Why can the mixed ether, methyl tert-butyl ether, be prepared using sulfuric acid catalyzed dehydration in this experiment?

（2）Why is diluted sulfuric acid used as the catalyst? What would happen if concentrated sulfuric acid was used?

（3）During the reaction, why is it necessary to strictly control the distillation temperature? What impact would a too fast distillation rate or too high distillation temperature have on the reaction?

3.8 Esterification Reaction

The esterification reaction refers to the reaction between carboxylic acids and alcohols or phenols under the catalysis of inorganic or organic strong acids, resulting in the formation of esters and water. Common catalysts include concentrated sulfuric acid, anhydrous hydrogen chloride, organic strong acids, or cation exchange resins. In esterification reactions, if the carboxylic acid itself is sufficiently acidic, such as formic acid or oxalic acid, no additional catalyst is required.

$$RCOOH+R'OH \xrightleftharpoons{H^+} RCOOR'+H_2O$$

The esterification reaction is a reversible reaction, and when the reaction reaches equilibrium, only about 65% of the acid and alcohol react to form esters. To favor the formation of esters, it is possible to continuously remove the product ester or water from the reactants, or to use an excess of carboxylic acid or alcohol. Whether to use an excess of acid or alcohol depends on factors such as the nature and price of the raw materials. For example, in the synthesis of ethyl acetate, excess ethanol is added to react with acetic acid because ethanol is cheaper than acetic acid. In addition, to remove the water generated in the reaction, azeotropic distillation is usually used. This involves adding some organic solvents, such as benzene, toluene, or chloroform, which can form azeotropes with water, and then distilling off the generated water. If the boiling point of the ester is lower than the boiling points of the acid, alcohol, and water, the method of continuously distilling off the ester can be used to shift the equilibrium forward. For example, this approach can be used in the synthesis of methyl formate and ethyl acetate.

3.8.1 General Method for Esterification Reaction

Set up an oil-water separator on a three-necked flask (if the boiling point of the ester is lower than that of the raw materials and water, there is no need to install an oil-water separator, and a distillation apparatus can be used directly to distill off the product ester during the reaction) and a reflux condenser. Add 0.1 mol carboxylic acid, 0.12 mol alcohol (or an excess of carboxylic acid) , 50 mL benzene, and 1 mL concentrated sulfuric acid to the reaction flask in sequence. Add a few zeolite crystals, heat and reflux. The vapor condenses through the condenser and flows into the oil-water separator, where the main components of the condensed liquid are benzene and water. When the benzene layer on the upper part accumulates to the outlet of the oil-water separator, it continuously flows back into the reaction flask. When the water layer on the lower part accumulates more, the water can be released by opening the piston of the oil-water separator.

After the reaction is complete, transfer the reaction mixture to a separatory funnel, add 40 mL water, shake and separate the lower water layer. Wash the remaining ester layer in the separatory funnel with 20 mL water, followed by 10 mL 5% sodium carbonate solution, and then wash it several times with water to make the organic phase neutral. Dry with anhydrous magnesium sulfate and then distill.

【Note】

(1) If the product ester itself can form an azeotrope with water, benzene or other organic solvents may not be added separately.

(2) Depending on the properties of the alcohol and ester, the benzene layer may also contain different proportions of alcohol and ester.

(3) The volume of collected water can be used to determine the end point of the reaction. However, the actual amount of water collected is higher than the theoretical calculated value because water is often distilled out in the form of azeotropes. Therefore, the volume of the corresponding azeotrope formed by an equal amount of water should be used to determine the end point. The end point can also be determined by observing whether there are water droplets continuing to sink in the oil-water separator.

(4) If it is difficult to separate the water layer and ester layer, saturated saline solution can be added for washing.

3.8.2 Experiment

Experiment Eight Acetylsalicylic Acid (Aspirin)

【Experimental objectives】

To learn the principle and experimental method of preparing esters from phenolic compounds as raw materials, and to consolidate the technique of recrystallization.

【Experimental principle】

$$C_6H_4(COOH)(OH) + CH_3COOCOCH_3 \xrightarrow{H_2SO_4} C_6H_4(COOH)(OCOCH_3) + CH_3COOH$$

Acetylsalicylic acid is white needle-like crystals, mp132～135℃ (Acetylsalicylic acid easily decomposes when heated, so the melting point is not obvious. During measurement, the bath liquid can be heated to about 110℃, and then the sample to be tested is placed in it for determination).

【Chemicals】

Salicylic acid	1.38 g (0.01 mol)
Acetic anhydride	4 mL (0.04 mol)
Concentrated sulfuric acid	a small amount
10% Sodium bicarbonate solution	20 mL
20% Hydrochloric acid	10 mL
1% Ferric chloride solution	a small amount

【Experimental procedure】

In a 100 mL conical flask, add 1.38 g of salicylic acid, 4 mL of acetic anhydride, and 4 drops of concentrated sulfuric acid in sequence. Shake well to dissolve the salicylic acid.

Note: Both acetic anhydride and concentrated sulfuric acid are highly corrosive. Be cautious when handling them. If they accidentally come into contact with the skin, rinse immediately with plenty of water.

Place the conical flask in a hot water bath at 60～70℃, heat for 10 minutes, and shake occasionally. Then stop heating and let the reaction mixture cool to room temperature. Slowly add 15 mL of water while shaking. Place the conical flask in a cold water bath for cooling, and crystals will precipitate. Filter and wash with a small amount of cold water, then dry to obtain crude acetylsalicylic acid.

Note: Due to the hydrolysis of the excess acetic anhydride, the reaction flask will become hot, and sometimes the reaction mixture may even boil. Be careful during the operation.

Transfer the crude product to a 100 mL beaker, add 10% sodium bicarbonate solution while stirring, until no more carbon dioxide is produced. Filter to remove insoluble polymers. Pour the filtrate into a 100 mL beaker, slowly add 10 mL of 20% hydrochloric acid while stirring, and crystals will gradually precipitate. Place the reaction mixture in an ice water bath to maximize crystal precipitation. Filter, wash with a small amount of cold water 2～3 times, then filter to dryness. Take a small amount of acetylsalicylic acid, dissolve it in a few drops of ethanol, and add 1～2 drops of 1% ferric chloride solution. If a color reaction occurs, the product can be recrystallized using an ethanol-water mixed solvent: first dissolve the crude product in a small amount of boiling ethanol, then add hot water to the ethanol solution until it becomes turbid, heat until the solution is clear, let it cool slowly, filter, dry, weigh, determine the melting point, and calculate the yield.

【Notes】

(1) Esterification of phenolic compounds is also called acylation, and commonly used acylating reagents include acyl chlorides and acid anhydrides. Compared with acyl chlorides, the reaction of acid anhydrides with phenolic compounds is milder.

(2) Adding a small amount of concentrated sulfuric acid or concentrated phosphoric acid accelerates the reaction.

(3) During the reaction, a small amount of salicylic acid undergoes a polymerization reaction to form a polymer. Aspirin can react with sodium bicarbonate to form a water-soluble salt, separating it from the polymer.

(4) Phenolic compounds can undergo a color reaction with ferric chloride solution, and this special color reaction can be used to test the presence of phenolic hydroxyl groups.

(5) During the recrystallization of acetylsalicylic acid, the solution should not be heated for too long, nor should high-boiling solvents be used, as acetylsalicylic acid is prone to decomposition at high temperatures.

【Thought questions】

(1) What is the role of concentrated sulfuric acid in the reaction between salicylic acid and acetic anhydride?

(2) Pure acetylsalicylic acid does not undergo a color reaction with ferric chloride solution. However, sometimes recrystallized acetylsalicylic acid in an ethanol-water mixed solvent may exhibit a color reaction with ferric chloride solution. What could be the reason for this?

(3) After the reaction between salicylic acid and acetic anhydride, if the method of separating and purifying the polymer impurity using sodium bicarbonate as a salt and hydrochloric acid for acidification is not employed, can you propose an alternative purification plan?

第四部分

有机化学基本实验

4.1 环己烯的制备

【实验目的】

掌握消除法制备烯烃，学习分馏及蒸馏操作。

【实验原理】

烯烃是重要的有机化工原料，工业上主要通过石油的裂解和催化脱氢来制备，在实验室主要通过醇脱水或卤代烃脱卤化氢来制备。

环己烯有特殊刺激性气味，不溶于水，溶于乙醇、醚。实验室常用环己醇作原料，在浓硫酸或浓磷酸催化作用下加热脱水来制备环己烯。本反应是可逆反应，采取边反应边蒸出产物（环己烯和水形成的二元共沸物，沸点 70.8℃，含水 10%）的方法，使得反应向正反应方向进行，以此获得较高收率。但是由于原料环己醇也能和水形成二元共沸物（沸点 97.8℃，含水 80%），为了保证使产物蒸出，而又不夹带原料，本实验采用分馏装置，并控制柱顶温度不超过 85℃。

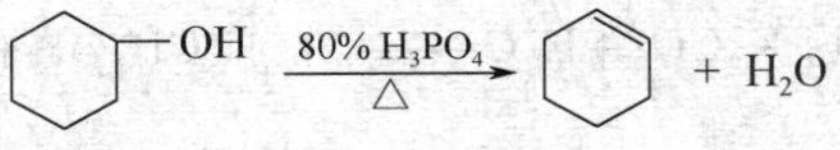

$$\text{C}_6\text{H}_{11}\text{—OH} \xrightarrow[\triangle]{80\%\ H_3PO_4} \text{C}_6\text{H}_{10} + H_2O$$

【药品】

环己醇	10 mL（9.6 g，约 0.1 mol）
磷酸（85%）	5 mL

饱和食盐水，无水氯化钙。

【实验所需时间】4 h

【实验步骤】

在 50 mL 干燥的圆底烧瓶中，放入 10 mL 环己醇及 5 mL 85%磷酸，充分摇荡使两种液体混合均匀。投入几粒沸石，参照图 2-10 安装分馏装置。用小锥形瓶作接收器，置于碎冰浴里。

用电热套缓慢加热反应物至沸腾，以较慢速度进行蒸馏并控制分馏柱顶部温度不超过73℃。当无液体蒸出时，提高电热套温度，继续蒸馏。当温度计读数到达85℃时，停止加热。蒸出液为环己烯和水的混浊液。

小锥形瓶中的粗产物，用吸管吸去水层，加入等体积的饱和食盐水，摇匀后静置待液体分层。用吸管吸去水层。油层转移到干燥的小锥形瓶中，加入少量粉末状无水氯化钙干燥。必须待液体完全澄清透明后，才能进行蒸馏。

将干燥后的粗制环己烯在水浴锅中进行蒸馏，收集82～85℃的馏分。所用的蒸馏装置必须是干燥的。

产量：4～5 g。

纯环己烯为无色透明液体，沸点83℃，d_4^{20} 0.8102，n_D^{20} 1.4465。

【注意事项】

（1）最好用油浴加热，使反应受热均匀。

（2）环己醇和水、环己烯和水皆形成二元恒沸混合物。

（3）粗环己烯也可以倒入小分液漏斗中进行后处理。

（4）蒸馏所得到的产物可以用气相色谱检测其纯度。固定液可用聚乙醇、邻苯二甲酸二壬酯等。

【思考题】

（1）用磷酸作脱水剂比用浓硫酸作脱水剂有什么优点？

（2）如果实验产率太低，试分析主要是在哪些操作步骤中造成损失。

4.2 苯甲醇的制备

【实验目的】

掌握卤代烃水解制备醇及搅拌装置的安装和使用。

【实验原理】

用苯氯甲烷制苯甲醇实质是卤代烃的水解反应。反应在碱性水溶液中进行。由于卤代烃一般不溶于水，故两相间（有机相和水相）反应比较缓慢，并且需要强烈搅拌。但如果加入相转移催化剂如四乙基溴化铵，这样就可以帮助反应物从一相转移到能够发生反应的另一相当中，从而有效提高反应速度，缩短反应时间。

$$2C_6H_5CH_2Cl + K_2CO_3 + H_2O \longrightarrow 2C_6H_5CH_2OH + 2KCl + CO_2$$

【药品】

苯氯甲烷	9.5 mL（10.1 g，0.08 mol）
碳酸钾	8 g（0.06 mol）
四乙基溴化铵	（50%水溶液）2 mL

无水硫酸镁，乙醚。

【实验所需时间】6 h

【实验步骤】

参照图2-6装配反应装置。

在装有机械搅拌器的250 mL三口烧瓶里加入碳酸钾水溶液（8 g碳酸钾溶于80 mL水中）及2 mL 50%的四乙基溴化铵水溶液，加几粒沸石。装上回流冷凝管和滴液漏斗，在滴液漏斗中加入 9.5 mL 苯氯甲烷。开动搅拌器，在电热套上加热至回流，将苯氯甲烷滴入三口烧瓶中。滴加完毕以后，继续在搅拌下加热回流，反应时间共 2 h。

停止加热，冷却到 30～40℃。把反应液移入分液漏斗中，分出油层。碱液用乙醚萃取 4 次，每次用 6 mL 乙醚。合并萃取液和粗苯甲醇。用无水硫酸镁或碳酸钾干燥。

将干燥透明的苯甲醇乙醚溶液倒入 50 mL 蒸馏烧瓶里，安装好蒸馏装置。先在热水浴上蒸出乙醚，然后改用空气冷凝管，在电热套上加热蒸馏。收集 200～208℃的馏分。

产量：约 5.5 g。

纯苯甲醇为无色透明液体，沸点 205.4℃，d_4^{20} 1.0419，n_D^{20} 1.5396。

【注意事项】

（1）也可用其他相转移催化剂，如三乙基苯甲基氯化铵。

（2）如不加相转移催化剂，反应需 6～8 h 才能完成。

（3）温度过低，碱会析出，给分离带来困难。

【思考题】

（1）在实验室中，还有哪些合适的方法可用来制备苯甲醇？

（2）本实验采用碳酸钾作为苯氯甲烷的碱性水解试剂，有何优点？

4.3　正丁醚的制备

【实验目的】

掌握低级伯醇脱水制备醚的方法，熟悉分水器的安装和使用。

【实验原理】

正丁醚是一种重要的有机溶剂。在酸催化下，醇分子间脱水是制备对称醚的常用方法。对于伯醇，一般可用此方法来合成醚类化合物；但对于仲醇和叔醇，由于它们比伯醇更容易发生消除反应，因此不适宜用此法来合成醚。

本实验利用正丁醇与浓硫酸反应来制备正丁醚。反应产物与温度的关系很大，在 90℃以下，正丁醇与硫酸失水生成硫酸酯。在较高温度（约 140℃）下，醇分子间脱水成醚。在更高温度（约 160 ℃）下，醇分子内脱水成烯。因此反应过程中须严格控制温度，以减少副产物的生成。

该反应是平衡反应，为了提高转化率，反应需在装有分水器的回流装置中进行。由于正丁醇的密度小于水，且在水中溶解度小，分水器可以使正丁醇不断返回到反应瓶中，而生成的水则沉于分水器的下端。间歇将反应中生成的水从反应体系中释放出来，可以达到油水分离的目的。

主反应：

$$2CH_3CH_2CH_2CH_2OH \xrightarrow{H_2SO_4} (CH_3CH_2CH_2CH_2)_2O + H_2O$$

副反应：

$$CH_3CH_2CH_2CH_2OH \xrightarrow{H_2SO_4} CH_3CH_2CH_2{=}CH_2 + H_2O$$

【药品】

正丁醇	31 mL（25 g，0.34 mol）
浓硫酸（d=1.84）	5 mL

50%硫酸，无水氯化钙。

【实验所需时间】6 h

【实验步骤】

在 100 mL 三口烧瓶中加入 31 mL 正丁醇，分批次加入 5 mL 浓硫酸并摇荡烧瓶使浓硫酸与正丁醇混合均匀，加几粒沸石。在烧瓶口上分别装上分水器和温度计，分水器上端再连一回流冷凝管。

分水器中可事先加入一定量的水（水的量等于分水器的总容量减去反应完全时可能生成的水量）。将烧瓶放在电热套中加热，保持回流 1 h。随着反应的进行，分水器中的水层不断增加，反应液的温度也逐渐上升。当分水器中的水层即将超过支管口而流回反应烧瓶时，可打开旋塞放掉少量水。当生成的水量到达 4.5～5 mL，瓶中反应液温度到达 150℃左右时，停止加热。如果加热时间过长，溶液会变黑并有大量副产物丁烯生成。

待反应物稍冷，拆除分水器，将仪器改装成蒸馏装置（参照图 2-9），加少量沸石，进行蒸馏，至无馏出液为止。

将馏出液倒入分液漏斗中，分去水层。粗产物用两份 15 mL 冷的 50%硫酸洗涤两次，再用水洗涤两次，最后用 1～2 g 无水氯化钙干燥。干燥后的粗产物倒入 30 mL 蒸馏烧瓶中（注意不要把氯化钙倒进去！）进行蒸馏（参照图 2-8），收集 140～144℃的馏分。

产量：7～8 g。

纯正丁醚为无色液体，沸点 142.4℃，d_4^{15} 0.773，n_D^{20} 1.3992。

【注意事项】

（1）本实验利用恒沸混合物蒸馏方法将反应生成的水不断从反应物中除去。正丁醇、正丁醚和水可能生成以下几种恒沸混合物：

恒沸混合物		沸点/℃	组成/%		
			正丁醚	正丁醇	水
二元	正丁醇-水	93.0		55.5	45.5
	正丁醚-水	94.1	66.6		33.4
	正丁醇-正丁醚	117.6	17.5	82.5	
三元	正丁醇-正丁醚-水	90.6	35.5	34.6	29.9

含水的恒沸混合物冷凝后分层，上层主要是正丁醇和正丁醚，下层主要是水。在反应过程中利用分水器使上层液体不断流回到反应器中。

（2）按反应式计算，生成水的量为 3 g。实际上分出水层的体积要略大于计算量，否则产率很低。

（3）也可以略去这一步蒸馏，而将冷的反应物倒入盛 50 mL 水的分液漏斗中，按下段的方法做下去。但因反应产物中杂物较多，在洗涤分层时有时会发生困难。

（4）50%硫酸可由 20 mL 浓硫酸与 34 mL 水配成。丁醇能溶于 50%硫酸中而正丁醚很少溶解。

【思考题】

如何得知反应已经完全？

4.4　正丁醛的制备

【实验目的】

掌握醇氧化制备醛的方法。

【实验原理】

醛的一个重要合成方法是通过醇氧化。工业上，比如甲醛的大量合成即通过氧气氧化甲醇获得。氧气属“绿色”试剂且廉价易得。实验室中则可使用更为多样的氧化剂，其中应用最普遍的有铬（Ⅵ）试剂。氧化剂可通过醇和酸性重铬酸钾溶液共热来制备，过量的重铬酸能氧化醛到羧酸形态。因此，需及时从反应体系中移除反应所生成的醛，或使用更温和的试剂，如吡啶重铬酸盐（PCC）制备醛，从而不用担心其过分氧化为酸。

主反应：

$$CH_3(CH_2)_2CH_2OH \xrightarrow[H_2SO_4]{Na_2CrO_7} CH_3(CH_2)_2CHO + H_2O$$

副反应：

$$CH_3(CH_2)_2CHO \xrightarrow[H_2SO_4]{Na_2CrO_7} CH_3(CH_2)_2COOH$$

$$2CH_3(CH_2)_2CH_2OH \xrightarrow[H_2SO_4]{Na_2CrO_7} CH_3(CH_2)_2COOC_4H_9 + H_2O$$

【药品】

正丁醇　　28 mL（22.2 g，0.3 mol）

重铬酸钠　　（$Na_2Cr_2O_7 \cdot 2H_2O$）30.5 g

浓硫酸　　（d=1.84）22 mL

无水硫酸镁或无水硫酸钠。

【实验所需时间】 6～8 h

【实验步骤】

仪器装置参照图 2-10（将烧瓶换成三口烧瓶）。

在 250 mL 烧杯，将 30.5 重铬酸钠溶解于 165 mL 水中。在仔细搅拌和冷却下，缓缓加入 22 mL 浓硫酸。将配制好的氧化剂倒入滴液漏斗中（可分数次加入）。往 250 mL 三口烧瓶里加入 28 mL 正丁醇及几粒沸石。

用电热套将正丁醇加热至微沸，待蒸气上升刚好达到分馏柱底部时，开始滴加氧化剂溶液，约在 20 min 内加完。注意滴加速度，使分馏柱顶部的温度不超过 78℃。同时，生成的正丁醛不断馏出。氧化反应是放热反应，在加料时要注意温度变化，控制柱顶温度不低于 71℃，又不高于 78℃。

当氧化剂全部加完后，继续反应约 15～20 min。收集所有在 95℃以下馏出的粗产物。

将此粗产物倒入分液漏斗中，分去水层。把上层的油状物倒入干燥的小锥形瓶中，加入 1～2 g 无水硫酸镁或无水硫酸钠进行干燥。

将澄清透明的粗产物倒入 30 mL 蒸馏烧瓶中，加入少量沸石。安装好蒸馏装置。在电热套中低温加热蒸馏，收集 70～80℃的馏出液。继续蒸馏，收集 80～120℃的馏分以回收正丁醇。

产量：约 7 g。

纯正丁醛为无色透明液体，沸点 75.7℃，d_4^{20} 0.817，n_D^{20} 1.3843。

【注意事项】

（1）正丁醛和水一起蒸出。接收瓶要用冰浴冷却。正丁醛和水形成二元恒沸混合物，其沸点为 68℃，恒沸物含正丁醛 90.3%。正丁醇和水也形成二元恒沸混合物，其沸点为 93℃，恒沸物含正丁醇 55.5%。

（2）绝大部分正丁醛应在 73～76℃馏出。正丁醛应保存在棕色的配磨口塞玻璃瓶内。

【思考题】

（1）制备正丁醛有哪些方法？

（2）为什么本实验中正丁醛的产率低？

（3）反应混合物颜色的变化说明什么？

（4）为什么采用无水硫酸镁或无水硫酸钠作干燥剂？

4.5 苯乙酮的制备

【实验目的】

（1）掌握傅-克（Friedel—Crafts）酰基化反应。

（2）掌握空气冷凝器的使用和减压蒸馏操作方法。

【实验原理】

Friedel-Crafts 酰基化反应是制备芳香酮的最重要和常用的方法之一，酸酐是常用的酰基化试剂，用无水 $FeCl_3$、BF_3、$ZnCl_2$ 和 $AlCl_3$ 等路易斯酸作催化剂。酰基化反应常用过量的芳烃、二硫化碳、硝基苯、二氯甲烷等作为反应的溶剂。这类反应一般为放热反应，通常是将酰基化试剂配成溶液后，慢慢滴加到盛有芳香族化合物的反应瓶中。

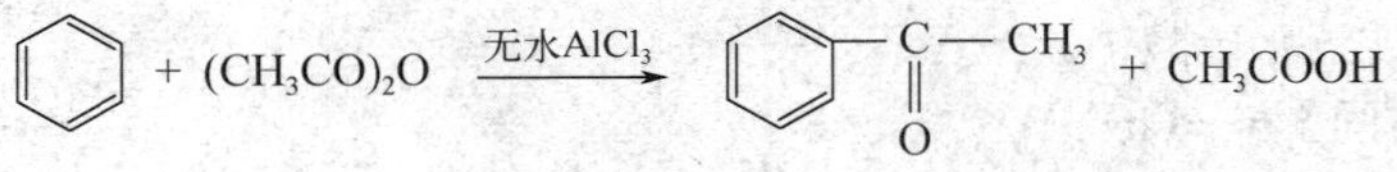

$$C_6H_6 + (CH_3CO)_2O \xrightarrow{\text{无水}AlCl_3} C_6H_5-\underset{\underset{O}{\|}}{C}-CH_3 + CH_3COOH$$

【药品】

苯	25 mL（22 g，0.282 mol）
无水三氯化铝	16 g（0.12 mol）
乙酸酐	4.7 mL（5.1 g，0.05 mol）

浓盐酸，浓硫酸，5%氢氧化钠溶液。

【实验所需时间】8 h

【实验步骤】

本实验所用的药品必须是无水的，所用的仪器必须是干燥的。

取 100 mL 三口烧瓶，在中间瓶口装配液封搅拌器，液封管内盛浓硫酸，一侧口装恒压滴液漏斗，另一侧口装回流冷凝管，回流冷凝管上口装上氯化钙干燥管并连接气体吸收装置（参照图 2-6 及图 2-7）。

在烧瓶中快速加入 16 g 无水三氯化铝和 20 mL 苯。在恒压滴液漏斗中放入 4.7 mL 新蒸馏过的乙酸酐和 5 mL 苯的混合液。在搅拌下慢慢滴加乙酸酐的苯溶液。反应很快开始，放出氯化氢气体，三氯化铝逐渐溶解，反应物的温度也自行升高。应控制滴加速度，使苯缓慢回流。加料时间约需 10 min。加完乙酸酐后，关闭恒压滴液漏斗旋塞，在电热套中加热，保持缓慢回流 1 h。

待反应物冷却后，在通风橱内把粗产物慢慢地倒入 50 g 碎冰中，同时不断搅拌。然后加入 30 mL 浓盐酸使析出的氢氧化铝沉淀溶解。如果仍有固体存在，再适当增加一点盐酸。用分液漏斗分出苯层。水层用 20 mL 苯分两次萃取。合并苯溶液，用 15 mL 5%氢氧化钠溶液洗涤，再用水洗涤。分出苯层。

在 30 mL 蒸馏烧瓶上装一个滴液漏斗，照 4.4 实验装配蒸馏装置。将吸收装置换成长橡皮管通入水槽或引至室外。将苯溶液倒入滴液漏斗中，先放约 10 mL 苯溶液到烧瓶中，在沸水浴上加热蒸馏，同时把剩余的苯溶液逐渐地滴加入烧瓶中，直至苯蒸不出为止（苯溶液中所含的少量水分随苯共沸蒸出）。卸去滴液漏斗，换上 250℃温度计，在电热套中加热蒸出残留的苯。当温度升至 140℃左右时，停止加热。稍冷后换空气冷凝管和接收器，继续蒸馏，收集 195～202℃的馏分。

产量：3.5～4 g。

纯苯乙酮是无色油状液体，熔点 19.6℃，沸点 202℃，d_4^{20} 1.028，n_D^{20} 1.5372。

【注意事项】

（1）本实验最好用无噻吩的苯。要除去苯中所含噻吩，可用浓硫酸多次洗涤（每次用相当于苯体积 15%的浓硫酸），直到不含噻吩为止，然后依次用水、10%氢氧化钠溶液和水洗涤，用无水氯化钙干燥后蒸馏。

（2）无水三氯化铝暴露在空气中，极易吸水分解而失效。应当用新升华过的或包装严密的试剂。称取时动作要迅速。块状的无水三氯化铝在称取前需在研体中迅速地研细。

加无水三氯化铝时可自制一简易的加料器：取一段长 6 cm，直径约 15 mm（可插入烧瓶的侧口）的玻璃管，两端配上橡皮塞。称量时装入药品，塞紧玻璃管两端。加料时，打开一塞，将玻璃管迅速插入瓶口，轻轻敲拍玻璃管使药品进入烧瓶，而不致粘在瓶口。如果有残留在管中的固体，可打开另一塞子，用玻璃棒将固体捅下去。

（3）仪器或药品不干燥，将严重影响实验结果或使反应难以进行。

（4）本实验也可用人工振荡代替机械搅拌。用 100 mL 圆底烧瓶，上装一个二口连接管，其正口装滴液漏斗，侧口装回流冷凝管，冷凝管上口连接氯化钙干燥管和气体吸收装置。为了便于振荡反应物质，烧瓶、冷凝管和滴液漏斗应安装在同一铁架台上。采用人工振荡时，回流时间应增长。

（5）回流时间增长，产率还可以提高。

（6）最好进行减压蒸馏，收集 86～90℃/1.6 kPa（12 mmHg）的馏分。

【思考题】

（1）为什么用过量的苯和无水三氯化铝？

（2）为什么缓慢加乙酸酐？

4.6　己二酸的制备

【实验目的】

学习氧化法制备酸和掌握重结晶基本操作。

【实验原理】

制备羧酸最常用的方法是烯、醇、醛等的氧化法。常用的氧化剂有硝酸、重铬酸钾

（钠）的硫酸溶液、高锰酸钾、过氧化氢及过氧乙酸等。

己二酸是合成尼龙-66以及聚氨酯的主要原料之一，实验室可用硝酸或高锰酸钾氧化环己醇来制备。使用硝酸时，会产生二氧化氮毒气，要注意通风或者尾气吸收。

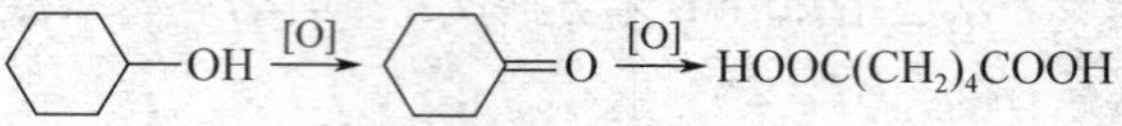

$$\text{C}_6\text{H}_{11}\text{OH} \xrightarrow{[O]} \text{C}_6\text{H}_{10}\text{=O} \xrightarrow{[O]} \text{HOOC(CH}_2\text{)}_4\text{COOH}$$

【药品】

环己醇	2.1 mL（2 g，0.02 mol）
硝酸（d=1.42）	5 mL（0.08 mol）

【实验所需时间】 2 h

【实验步骤】

本实验必须在通风橱内进行。做实验时必须严格地遵照规定的反应条件。

在50 mL圆底烧瓶中放一支温度计，其水银球要尽量接近瓶底。用有直沟的单孔软木塞将温度计夹在铁架上。

在烧瓶中加5 mL水，再加5 mL硝酸。将溶液混合均匀，在水浴上加热到80℃，然后用滴管加2滴环己醇。反应立即开始。温度随即上升到85～90℃。小心地逐渐滴加2.1 mL环己醇，一定要使温度维持在这个范围内，必要时往水浴中添加冷水。当醇全部加入而且溶液温度降低到80℃以下时，将混合物在85～90℃下加热2～3 min。

在冰浴中冷却，析出的晶体在布氏漏斗上进行抽滤。用滤液洗出烧瓶中剩余的晶体。用3 mL冰水洗涤己二酸晶体，抽滤。晶体再用3 mL冰水洗涤一次，再抽滤。取出产物，晾干。

产量：约1.4 g。

纯己二酸是无色单斜晶体，熔点153℃。

4.7 乙酸正丁酯的制备

【实验目的】

掌握基本酯化反应。

【实验原理】

羧酸酯是用途很广的一类有机化合物。许多酯是调配食品或化妆品香精的原料。油脂不仅可作为食用油，而且是重要的工业原料。邻苯二甲酸二丁酯和二辛酯是人造革及聚氯乙烯的增塑剂。除虫菊酯及其类似物是一类高效、低毒、残留少的农药。

羧酸酯常用羧酸与醇在酸催化下直接酯化来制备。常用的催化剂有硫酸、干燥的氯化氢、对甲苯磺酸、强酸性树脂等。

$$CH_3COOH + n\text{-}C_4H_9OH \rightleftharpoons CH_3COOC_4H_9\text{-}n + H_2O$$

该反应是可逆的，如果使用等摩尔的羧酸和醇反应，达到平衡时，只有2/3左右的原料转化为酯。为了提高产率，一般使一种原料大量过量（根据酸和醇的来源、价格及是否容易与产物分离等因素来确定哪一种原料过量）。除去反应中生成的水也是提高收率的常用方法。在某些酯化反应中，醇、酯和水可以形成二元或三元恒沸物，可在反应体系中加入与水能形成共沸物的溶剂（如苯），利用分水器可以达到满意的效果。

实验证明，在反应体系中加入的催化剂过量，可以加大反应速率，由于酸与水的水合作用可以除去反应中生成的部分水，从而提高酯的产量。

【药品】

正丁醇　　　　　　　　11.5 mL（9.3 g，0.125 mol）

冰醋酸　　　　　　　　7.2 mL（7.5 g，0.125 mol）

浓硫酸，10%碳酸钠溶液，无水硫酸镁。

【实验所需时间】4 h

【实验步骤】

在干燥的 50 mL 圆底烧瓶中，加入 11.5 mL 正丁醇和 7.2 mL 冰醋酸，再加入 3～4 滴浓硫酸。混合均匀，加入少量沸石，然后安装分水器及回流冷凝管，并在分水器中预先加水至略低于支管口（记录加入的水的体积）。在电热套中加热回流，反应一段时间后把水逐渐分去，保持分水器中水层液面在原来的高度。约 40 min 后不再有水生成，表示反应完毕。停止加热，记录分出的水量。冷却后卸下回流冷凝管，把分水器中分出的酯层和圆底烧瓶中的反应液都转移到分液漏斗中。用 10 mL 水洗涤，分去水层。酯层用 10 mL 10%碳酸钠溶液洗涤，检验是否呈酸性（如仍有酸性怎么办？）。分去水层。将酯层再用 10 mL 水洗涤一次，分去水层。将酯层倒入小锥形瓶中，加少量无水硫酸镁干燥。

将干燥后的乙酸正丁酯倒入干燥的 30 mL 蒸馏烧瓶中（注意不要把硫酸镁倒进去!），加入沸石，安装好蒸馏装置，在电热套中加热蒸馏。收集 124～126℃的馏分。前后馏分倒入指定的回收瓶中。

产量：10～11 g。

纯乙酸正丁酯是无色液体，沸点 126.5℃，d_4^{20} 0.882，n_D^{20} 1.3951

【注意事项】

本实验利用恒沸混合物蒸馏方法除去酯化反应中的水，下层为水层。

恒沸混合物		沸点/℃	组成/%		
			乙酸正丁酯	正丁醇	水
二元	乙酸正丁酯-水	90.7	72.9		27.1
	正丁醇-水	93.0		55.5	44.5
	乙酸正丁酯-正丁醇	117.6	32.8	67.2	
三元	乙酸正丁酯-正丁醇-水	90.7	63.0	34.6	29.0

4.8 3-丁酮酸乙酯的制备

【实验目的】

学习金属钠的操作和巩固减压蒸馏基本操作。

【实验原理】

含 α-活泼氢的酯在强碱性试剂（如 $NaNH_2$、NaH、$NaOCH_2CH_3$ 或三苯甲基钠）存在下，能与另一分子酯发生克莱森（Claisen）酯缩合反应，生成 β-羰基酸酯。3-丁酮酸乙酯就是通过这一反应制备的。

通常以酯及金属钠为原料，并以过量的酯作为溶剂，利用酯中含有的微量醇与金属钠反应来生成醇钠。由于醇的不断生成，反应能不断地进行下去，直至金属钠消耗完毕。

但作为原料的酯中含醇量过高又会影响到产品的得率，故一般要求中含醇量在 3%以下。

所制得的 3-丁酮酸乙酯是一个酮式和烯醇式混合物，在室温下含有 93%的酮式及 7%的烯醇式。

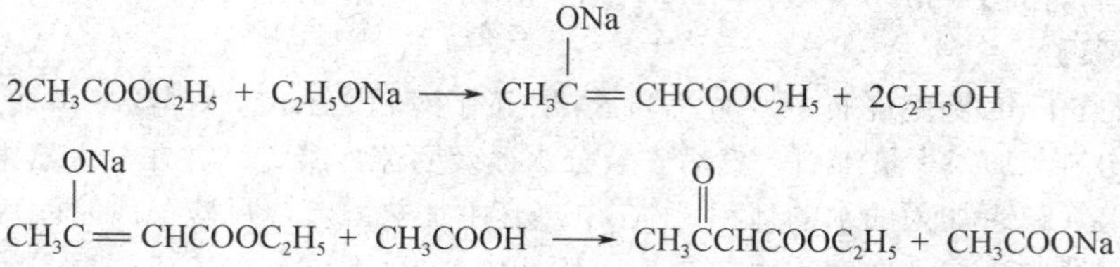

$$2CH_3COOC_2H_5 + C_2H_5ONa \longrightarrow CH_3\overset{\displaystyle ONa}{\overset{|}{C}}=CHCOOC_2H_5 + 2C_2H_5OH$$

$$CH_3\overset{\displaystyle ONa}{\overset{|}{C}}=CHCOOC_2H_5 + CH_3COOH \longrightarrow CH_3\overset{\displaystyle O}{\overset{\|}{C}}CHCOOC_2H_5 + CH_3COONa$$

【药品】

乙酸乙酯　　48.9 mL（44 g，0.5 mol）

金属钠　　5 g（约 0.22 mol）

稀醋酸，5%碳酸钠溶液，无水碳酸钾，饱和食盐水，饱和氯化钙溶液。

【实验所需时间】8 h

【实验步骤】

本实验所用的药品必须是无水的，所用的仪器必须是干燥的。

在干燥的 250 mL 圆底烧瓶中，放入 48.9 mL 无水的乙酸乙酯和切细的 5 g 金属钠。迅速装上回流冷凝管，其上口连接一个装有氯化钙的干燥管，如图 2-4 所示。用水浴加热，促使反应开始。若反应过于剧烈，可暂时移去热水浴而用冷水浴冷却。待反应和缓后，再用水浴加热，保持缓缓回流。待金属钠全部反应完后，停止加热。这时反应混合物变为红色透明并呈绿色荧光的液体（有时析出黄白色沉淀）。冷却至室温，卸下冷凝管。将烧瓶浸在冷水浴中，在摇动下缓慢地滴加稀醋酸，使呈弱酸性（用什么试纸检验？），这时所有固体物质都溶解。用分液漏斗分离出红色的酯层。用 20 mL 乙酸乙酯提取水层中的酯，并入原酯层。酯层用 5%碳酸钠溶液洗至中性，再用无水碳酸钾或无水硫酸镁干燥。

将干燥的液体倒入 125 mL 克氏蒸馏烧瓶内。装配好减压蒸馏装置（装配和操作方法见 2.5）。先在常压下蒸出乙酸乙酯。然后在减压下蒸出 3-丁酮酸乙酯。所收集的馏分的沸点范围视压力而定：

压力/kPa	1.666	1.866	2.399	3.866	5.998	10.66
压力/mmHg	12.5	14	18	29	45	80
沸点/℃	71	74	79	88	94	100
沸点范围/℃	69～73	72～76	77～81	86～90	92～96	98～102

产量：8～9.5 g。

纯 3-丁酮酸乙酯为无色液体，沸点 180℃/0.1006MPa（755 mmHg）（蒸馏时稍有分解），d_4^{20} 1.025，n_D^{20} 1.4194。

【注意事项】

（1）所用的乙酸乙酯必须是无水的，其提纯方法如下：用饱和氯化钙溶液将普通的乙酸乙酯洗涤数次，再用熔融过的无水碳酸钾干燥，在水浴上蒸馏，收集 76～77℃的馏分。

（2）金属钠颗粒的大小直接影响缩合反应的速率。用镊子从瓶中取出金属钠块，用双层滤纸吸干溶剂油，再用刀切去金属钠表面的氧化层。迅速称量，立即用金属钠压丝机直接压入烧瓶内（或压入无水乙醚中）。如无金属钠压丝机，则可（在冰冷的无水乙醚中）将金属钠切成细条，再立刻移入盛无水乙酸乙酯的烧瓶中（注意尽量缩短金属钠与空气接触的时间）。

（3）金属钠全部消失所需时间视钠的颗粒大小而定，一般需 1.5～3 h。

（4）缩合反应这一步骤必须在一次实验课内完成，否则会影响产量。

（5）滴加稀醋酸时，需特别小心，如果反应物内杂有少量未反应完的金属钠，会发生剧烈反应。在此操作中还应避免加入过量的醋酸溶液，否则将会增加酯在水中的溶解度，降低产量。室温高时可用 30%醋酸，室温低时可用 25%醋酸。

（6）3-丁酮酸乙酯的互变异构现象可通过以下简单的试验来观察：取 2～3 滴制成的 3-丁酮酸乙酯溶于 2 mL 水中，加 1 滴 1%三氯化铁溶液，观察溶液颜色的变化。再很快地滴加溴水至溶液的颜色褪去为止。静置并观察颜色的变化。颜色重新显出后，可再滴加溴水，多次重复上述实验。

【思考题】

（1）所用仪器未经干燥处理，对反应有什么影响？为什么？

（2）为什么最后一步要用减压蒸馏法？

（3）描述 3-丁酮酸乙酯的互变异构现象。

4.9　环己酮的制备

【实验目的】

学习简易的水蒸气蒸馏的操作。

【实验原理】

环己酮属于脂环酮，具有类似丙酮的气味，常温下为无色或淡黄色、透明的油状液体，微溶于水，可溶于各种有机溶剂。环己酮是制备己内酰胺、己二酸等的原料。

实验室常通过伯醇或仲醇的氧化来制备相应的醛或酮。环己醇是仲醇，能被酸性的重铬酸钠（或钾）氧化成环己酮。酮的稳定性较高，一般不易被进一步氧化。但必须严格控制好反应条件，勿使氧化反应进行得过于剧烈，否则产物会进一步被氧化而发生碳链的断裂。

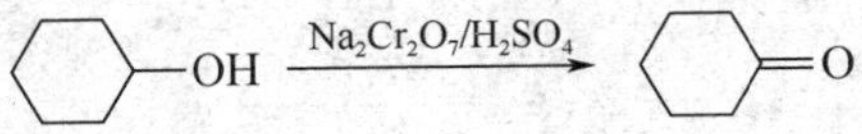

$$\text{C}_6\text{H}_{11}\text{—OH} \xrightarrow{Na_2Cr_2O_7/H_2SO_4} \text{C}_6\text{H}_{10}\text{=O}$$

【药品】

环己醇	10.4 mL（10 g，0.1 mol）
重铬酸钠	（$Na_2Cr_2O_7 \cdot 2H_2O$）10.4 g（0.035 mol）
浓硫酸（d=1.84）	10 mL

甲醇，精盐，无水硫酸镁。

【实验所需时间】 6 h

【实验步骤】

在 250 mL 圆底烧瓶内，放入 60 mL 冰水，一边摇动烧瓶，一边慢慢地加入 10 mL 浓硫酸，再小心地加入 10.4 mL 环己醇。将溶液冷却至 15℃。

在 100 mL 烧杯内，将 10.4 g 重铬酸钠水合物溶于 10 mL 水中。将此溶液冷却到 15℃，并分几批加到环己醇的硫酸溶液中。要不断地摇动烧瓶，使反应物充分混合。第一批重铬酸钠溶液加入后，不久反应物温度自行上升，反应物由橙红色变成墨绿色。反应物温度升到 55℃时，可用冷水浴适当冷却，控制反应温度在 55～60℃。待反应物的橙红色完全消失后，方可加下一批。待重铬酸钠溶液全部加完后，继续摇动烧瓶，直至反应温度出现下降趋势。再间歇摇动 5～10 min。之后加入 1～2 mL 甲醇以还原过量的氧化剂。

在反应物内加入 50 mL 水及沸石，安装成蒸馏装置。在电热套中加热蒸馏，把环己酮和水一起蒸出来，收集约 40 mL 馏出液。馏出液中加入约 8 g 精盐，搅拌促使食盐溶解。将此液体移入分液漏斗中，静置。分离出有机层（环己酮），用无水硫酸镁干燥。蒸馏，收集 151～156℃的馏分。

产量：约 6 g。

纯环己酮为无色液体，沸点 155.7℃，d_4^{20} 0.948，n_D^{20} 1.4507。

【注意事项】

（1）反应物不宜过于冷却，以免积累起未反应的铬酸。当铬酸达到一定浓度时，氧化反应会进行得非常剧烈，有失控的危险。

（2）甲醇也可以用 0.5～1 g 草酸代替。

（3）这步蒸馏操作，实质上是一种简化了的水蒸气蒸馏。环己酮和水形成恒沸混合物，沸点 95℃，含环己酮 38.4%。

（4）31℃时，环己酮在水中的溶解度为 2.4 g/100 g 水。馏出液中加入食盐是为了降低环己酮的溶解度，并有利于环己酮的分层。

【思考题】

（1）在加重铬酸钠溶液过程中，为什么要待反应物的橙红色完全消失后，方能加下一批重铬酸钠？在整个氧化反应过程中，为什么要控制温度在一定的范围？

（2）氧化反应结束后，为什么要往反应物中加入甲醇或草酸？

（3）如果从反应混合液中蒸馏出过多的馏出液，会有什么结果？如何弥补？

（4）从反应混合物中分离出环己酮，除了现在采用的水蒸气蒸馏法外，还可采用何种方法？

（5）在蒸馏环己酮、收集 151～156℃的馏分时，应选用水冷却型冷凝管还是空气冷凝管？

Basic Experiments

4.1 Preparation of cyclohexene

【Experimental objectives】

To learn how to synthesize alkenes by means of elimination and the techniques of fractional distillation and distillation.

【Experimental principle】

Alkenes are important organic chemical raw materials, primarily prepared industrially through petroleum cracking and catalytic dehydrogenation. In the laboratory, alkenes are mainly prepared through alcohol dehydration or halogenated hydrocarbon dehydrohalogenation.

Cyclohexene has a distinct irritant odor and is insoluble in water but soluble in ethanol and ether. In the laboratory, cyclohexanol is commonly used as a starting material, and cyclohexene is prepared by heating with concentrated sulfuric acid or concentrated phosphoric acid as catalysts for dehydration. This reaction is reversible, and to achieve higher yields, a method of simultaneously removing the products（cyclohexene and water, which form an azeotrope with a boiling point of 70.8℃and 10% water）during the reaction is employed to drive the reaction in the forward direction. However, since the starting material cyclohexanol can also form an azeotrope with water（boiling point 97.8℃ and 80% water）, a distillation apparatus is used in this experiment to ensure the removal of the product without carrying over the starting material cyclohexanol, and the column top temperature is controlled not to exceed 85℃.

$$\text{C}_6\text{H}_{11}\text{OH} \xrightarrow[\Delta]{80\%\ H_3PO_4} \text{C}_6\text{H}_{10} + H_2O$$

【Chemicals】

Cyclohexanol	10 mL（9.6 g, about 0.1 mol）
Phosphoric acid	5 mL（85%）

Saturated aqueous sodium chloride solution, Anhydrous calcium chloride.

【Time】 6 h

【Experimental procedure】

Mix 10 mL of cyclohexanol and 5 mL of 85% phosphoric acid in a 50 mL dry RBF. Zeolite is added and a fractional distillation device is installed. A small conical flask is put in an ice bath as an adaptor.

Heat the mixture gently to reflux, control the temperature of the outlet of the fractional distillation column not to exceed 73°C until no liquid is distilled out. Keep heating until the temperature reaches 85°C and stop it. A mixture of cyclohexene and water is obtained as a distillate.

The crude product was treated by removing water with a dropper, the same volume of saturated salt solution as the organic layer was added. Shaking, remove the aqueous layer with a dropper, transfer the organic layer to a dry conical flask. The moisture inside is removed by adding anhydrous calcium chloride.

Distil the clear liquid and collect the parts between 82～85°C. Note: all the apparatuses are dry.

Yield: 4～5 g.

Data: colorless liquid, bp 83°C, d_4^{20} 0.8102, n_D^{20} 1.4465.

【Notes】

(1) It is preferable to heat the reaction using an oil bath to ensure even heating.

(2) Cyclohexanol and water, as well as cyclohexene and water, both form a binary azeotropic mixture.

(3) The crude cyclohexene can also be poured into a small separatory funnel for post-treatment.

(4) The product obtained from distillation can be analyzed for purity by means of gas chromatography. Common stationary phases include polyethylene glycol and dioctyl phthalate, etc.

【Thought questions】

(1) Are there any advantages to phosphoric acid as a dehydrating agent over concentrated sulfuric acid?

(2) Please analyze the losses existing in your experimental procedure if your yield is very low.

4.2 Preparation of benzyl alcohol

【Experimental objectives】

To learn how to synthesize alcohols by the hydrolysis of halogenated hydrocarbons

【Experimental principle】

The synthesis of benzyl alcohol from benzyl chloride is essentially a hydrolysis reaction of the halogenated hydrocarbon. The reaction takes place in an alkaline aqueous solution. Since halogenated hydrocarbons are generally insoluble in water, the reaction between the two phases (organic phase and aqueous phase) is relatively slow and requires vigorous stirring. However, if a phase transfer catalyst such as tetraethylammonium bromide is added, it helps the reactants transfer from one phase to the other phase where the reaction can occur, thereby effectively increasing the reaction rate and shortening the reaction time.

$$2C_6H_5CH_2Cl + K_2CO_3 + H_2O \longrightarrow 2C_6H_5CH_2OH + 2KCl + CO_2$$

【Chemicals】

phenyl chloromethane 9.5 mL (10.1 g, 0.08 mol)
potassium carbonate 8 g (0.06 mol)
50% tetraethyl ammonium bromide 2 mL
anhydrous $MgSO_4$, ether.

【Time】 6 h

【Experimental procedure】

Install the device according to Scheme 2-6.

In a 250 mL three neck RBF, a mixture of aqueous potassium carbonate (8 g/80 mL) and 2 mL of 50% tetraethyl ammonium bromide is added. Zeolite is added. 9.5 mL of phenyl chloromethane is added into the dropping funnel, stir the mixture and heat to reflux. Phenyl chloromethane is added dropwise within 10 min. and continue to reflux for 2 h afterwards.

Stop heating, cool the mixture to 30～40°C. Transfer the mixture to a separatory funnel and collect the organic layer. The aqueous layer is extracted four times by ether, 6 mL each time. Combine them with the crude benzyl alcohol, dry the mixture with anhydrous magnesium sulfate or potassium carbonate.

Distill the crude benzyl alcohol in a 50 mL RBF. At first, distill ether with warm heating, then change the water condenser to an air condenser, continue to distill and collect the part of 200～208°C.

Yield: about 5.5 g.

Data: colorless liquid, bp 205.4°C, d_4^{20} 1.0419, n_D^{20} 1.5396.

【Notes】

(1) Other phase transfer catalysts, such as triethylbenzylammonium chloride, can also be used.

(2) Without the addition of a phase transfer catalyst, the reaction takes 6-8 hours to complete.

(3) If the temperature is too low, the alkali will precipitate, causing difficulties in separation.

【Thought questions】

(1) In a laboratory, how many convenient methods can be used to synthesize benzyl alcohol other than the method described here?

(2) What is the advantage of potassium carbonate functioning as a basic hydrolytic reagent of phenyl chloromethane?

4.3 Preparation of dibutyl (*n*–) ether

【Experimental objectives】

To learn the preparation of ether by means of alcohol dehydration. Be familiar with the installation and usage of water knockout vessel.

【Experimental principle】

n-Butyl ether is an important organic solvent. Under acid catalysis, the dehydration between alcohol molecules, is a common method for synthesizing symmetrical ethers. This method can be used to synthesize ether compounds from primary alcohols. However, for secondary alcohols and tertiary alcohols, they are more prone to undergo elimination reactions than primary alcohols, making this method unsuitable for synthesizing ethers.

This experiment uses the reaction between *n*-butanol and concentrated sulfuric acid to prepare dibutyl（*n*-）ether. The relationship between the reaction product and temperature is significant. Below 90°C, *n*-butanol reacts with sulfuric acid to form sulfate esters through dehydration. At higher temperatures（around 140°C）, dehydration between alcohol molecules occurs, forming ethers. At even higher temperatures（around 160°C）, intra-molecular dehydration of alcohol molecules leads to the formation of alkenes. Therefore, strict temperature control is necessary during the reaction to minimize the formation of by-products.

The reaction is an equilibrium reaction. To improve the conversion rate, the reaction needs to be carried out in a reflux apparatus equipped with a separator. Since the density of *n*-butanol is lower than that of water and its solubility in water is small, the separator allows *n*-butanol to continuously return to the reaction flask while the generated water settles at the bottom of the separator. Periodically releasing the water produced from the reaction system can achieve the purpose of oil-water separation.

Main reaction:

$$2CH_3CH_2CH_2CH_2OH \xrightarrow{H_2SO_4} (CH_3CH_2CH_2CH_2)_2O + H_2O$$

Side Reaction:

$$CH_3CH_2CH_2CH_2OH \xrightarrow{H_2SO_4} CH_3CH_2CH_2=CH_2 + H_2O$$

【Chemicals】

n-butanol	31 mL（25 g, 0.34 mol）
H_2SO_4（*d*=1.84）	5 mL

50% H_2SO_4, anhydrous calcium chloride.

【Time】 6 h

【Experimental procedure】

Add 31 mL of *n*-butanol into a 100 mL RBF, then 5 mL of concentrated sulfuric acid is added slowly into the above RBF and shake. Zeolite is added. Water knockout vessel and a thermometer are installed around the neck of the RBF, a refluxing condenser is also installed along with the water knockout vessel.

A definite volume of water is added into the water knockout vessel（the amount of water is equal to the total capacity of the water knockout vessel minus the water that might be produced if the reaction is complete), heat the mixture gently and keep refluxing for 1 h. With the progress of the reaction, the water volume in the water knockout vessel has increased together with the increasing temperatures. If the water inside the knockout vessel is full, release some water. When the reacting temperature reaches 150°C, stop heating. Longer reaction time will lead to a lot of side reactions together with a black residue.

Cool down the mixture and discharge the apparatus and switch to a distillation device. Zeolite is added, start to distill until no liquid is distilled out.

The distillate is poured into a separatory funnel. Aqueous layer is removed. The crude product is washed with 15 mL of cold 50% sulfuric acid twice. Afterwards, wash with water and then dry with anhydrous calcium chloride.

The crude product is distilled to collect the part between 140～144°C.

Yield: 7～8 g.

Data: colorless liquid, bp 142.4°C, d_4^{15} 0.773, n_D^{20} 1.3992.

【Notes】

(1) This experiment utilizes the method of azeotropic distillation to continuously remove the water produced from the reactants. The azeotropic mixtures that may be formed include *n*-butanol/water, dibutyl (*n*-) ether/water, and *n*-butanol/dibutyl (*n*-) ether/water.

Azeotropic Mixture		bp/°C	Composition/%		
			dibutyl (*n*-) ether	*n*-butanol	water
Binary	*n*-butanol/water	93.0		55.5	45.5
	dibutyl (*n*-) ether/water	94.1	66.6		33.4
	n-butanol/dibutyl (*n*-) ether	117.6	17.5	82.5	
Ternary	*n*-butanol/dibutyl (*n*-) ether/water	90.6	35.5	34.6	29.9

The water-containing azeotropic mixture is condensed and separated into two layers. The upper layer mainly consists of *n*-butanol and dibutyl (*n*-) ether, while the lower layer mainly consists of water. During the reaction process, a separatory funnel is used to continuously return the upper layer of liquid back to the reaction vessel.

(2) According to the stoichiometric calculation, the amount of water generated is 3 g. In practice, the volume of the separated water layer should be slightly larger than the calculated amount; otherwise, the yield will be low.

(3) Alternatively, this distillation step can be skipped, and the cooled reactants can be poured into a separatory funnel containing 50 mL of water. The subsequent steps can be carried out using the method described below. However, due to the presence of impurities in the reaction products, difficulties may arise during the washing and separation process.

(4) A mixture of 50% sulfuric acid can be prepared by combining 20 mL of concentrated sulfuric acid with 34 mL of water. *n*-Butanol is soluble in 50% sulfuric acid, whereas dibutyl (*n*-) ether dissolves only to a small extent.

【Thought questions】

How to tell the reaction is complete?

4.4 Preparation of *n*-butyraldehyde

【Experimental objectives】

To learn the preparation of aldehyde from the oxidation of alcohols.

【Experimental principle】

One important method for synthesizing aldehydes is through alcohol oxidation. In industry, for example, a large amount of formaldehyde is synthesized by oxidizing methanol with oxygen. Oxygen is considered a "green" reagent and is inexpensive and easily accessible. In the laboratory, a variety of oxidizing agents can be used, with chromium（Ⅵ）reagents being the most common. The oxidation reaction can be carried out by heating the alcohol with an acidic solution of potassium dichromate. Excess potassium dichromate can oxidize the aldehyde to its carboxylic acid form. Therefore, it is necessary to promptly remove the aldehyde generated from the reaction system or use milder reagents, such as pyridinium chlorochromate（PCC）, to prepare the aldehyde, thereby avoiding excessive oxidation to the acid form.

Main reaction:

$$CH_3(CH_2)_2CH_2OH \xrightarrow[H_2SO_4]{Na_2CrO_7} CH_3(CH_2)_2CHO + H_2O$$

Side reactions:

$$CH_3(CH_2)_2CHO \xrightarrow[H_2SO_4]{Na_2CrO_7} CH_3(CH_2)_2COOH$$

$$2CH_3(CH_2)_2CH_2OH \xrightarrow[H_2SO_4]{Na_2CrO_7} CH_3(CH_2)_2COOC_4H_9 + H_2O$$

【Chemicals】

n-butanol	28 mL（22.2 g, 0.3 mol）
Sodium dichromate（$Na_2Cr_2O_7 \cdot 2H_2O$）	30.5 g
Concentrated sulfuric acid	22 mL（*d*=1.84）

Anhydrous magnesium sulfate or sodium sulfate.

【Time】 6～8 h

【Experimental procedure】

Install the apparatus according to Scheme 2-10（three-neck rbf is used instead）.

In a 250 mL beaker, dissolve sodium dichromate 30.5 g in 165 mL water and stir gently. 22 mL of concentrated sulfuric acid is added slowly. The oxidant is poured into a dropping funnel. Add 28 mL of *n*-butanol and several pieces of zeolite into the three-neck rbf.

Heat the liquid to boil gently. Once the temperature increases, add the oxidant solution dropwise. Control the temperature of the thermometer between 71°C and 78°C.

When the oxidant solution is added completely, keep heating gently for another 15～20 minutes. Collect crude product below 95°C.

Pour the crude product into a separatory funnel and the aqueous layer is removed. Pour the upper organic layer into a dry conical flask, the moisture is removed by adding anhydrous magnesium sulfate or anhydrous sodium sulfate.

Pour the clear crude product into a 30 mL distillation flask zeolites are added. Install the distillation device and start to distill, collect the parts of 70～80℃. Keep heating, collect the part of 80～120℃ to recycle *n*-butanol.

Yield: 7 g.

Data: colorless clear liquid, bp 75.7℃, d_4^{20} 0.817, n_D^{20} 1.3843.

【Note】

(1) *n*-Butyraldehyde is distilled together with water. The receiving flask should be cooled in an ice bath. *n*-Butyraldehyde and water form an azeotropic mixture with a boiling point of 68℃, and the azeotrope contains 90.3% *n*-butyraldehyde. *n*-Butanol and water also form an azeotropic mixture with a boiling point of 93℃, and the azeotrope contains 55.5% *n*-butanol.

(2) The majority of *n*-butyraldehyde should be distilled at 73～76℃. *n*-Butyraldehyde should be stored in a brown glass stoppered bottle.

【Thought questions】

(1) How many methods can be used to synthesize *n*-butanal?

(2) Why is the reaction yield very low?

(3) What does the change in color of the reaction mixture tell us?

(4) Why use anhydrous $MgSO_4$ or Na_2SO_4 as desiccant?

4.5 Preparation of acetophenone

【Experimental objectives】

(1) To learn Friedel-Crafts Acylation Reaction.

(2) To learn the usage of air condensers and the operation of vacuum distillation.

【Experimental principle】

The Friedel-Crafts acylation reaction is one of the most important and commonly used methods for preparing aromatic ketones. Acid anhydrides are commonly used as acylating agents, and Lewis acids such as anhydrous $FeCl_3$, BF_3, $ZnCl_2$, and $AlCl_3$ are used as catalysts. Excess liquid aromatic hydrocarbons, carbon disulfide, nitrobenzene, dichloromethane, etc., are commonly used solvents in the acylation reaction. These reactions are generally exothermic and typically involve slowly adding the acylating agent solution to a reaction flask containing aromatic compounds.

【Reaction】

$$C_6H_6 + (CH_3CO)_2O \xrightarrow{\text{Anhydrous aluminium chloride}} C_6H_5\text{—}\overset{\overset{\displaystyle O}{\|}}{C}\text{—}CH_3 + CH_3COOH$$

【Chemicals】

Benzene	25 mL (22 g, 0.282 mol)
Anhydrous aluminium chloride	16 g (0.12 mol)
Acetic anhydride	4.7 mL (5.1 g, 0.05 mol)

Concentrated hydrochloric acid, Concentrated sulfuric acid, 5% sodium hydroxide solution.

【Time】 8 h

【Experimental procedure】

All the chemicals used are anhydrous, the apparatuses must be dry.

Install the device according to Scheme 2-6 and 2-7, the $CaCl_2$ drying tube is connected to the refluxing condenser, on the other side of the drying tube, the gas absorption device is connected.

16 g anhydrous aluminium chloride and 20 mL benzene are added into the flask as soon as possible. A mixture of 4.7 mL fresh acetic anhydride and 5 mL benzene are added into the dropping funnel. The mixture is added by dropwise into the flask while stirring. The reaction set in immediately and hydrochloric gas evolved. Keep the addition speed to let the mixture reflux gently. The process takes about 10 minutes. Then keep heating and let the mixture reflux gently for one hour.

When the reactants have cooled down to room temperature, pour the mixture into 50 g chopped ice inside the ventilating cabinet. Keep stirring. Add 30 mL concentrated hydrochloric acid to dissolve the solid inside the mixture. Separate the benzene layer with a separatory funnel. The aqueous layer is extracted with benzene three parts of benzene layer are combined and are washed with 15 mL of 15% NaOH soln., then washed with water. Collect the benzene layer.

Install a distillation device. Put a dropping funnel around the neck. Add 10 mL benzene inside the flask first, then heat the mixture to distill, in the meantime, add the mixture inside the dropping funnel dropwise. Keep distilling until no benzene is out. Remove the dropping funnel and put a thermometer（250℃）instead. Keep heating to distill the remaining benzene. When the temperature reaches 140℃, stop heating, replace the water-condenser with an air condenser. Keep heating and collect the part between 195～202℃.

Yield: 3.5～4 g.

Data: colorless liquid, mp. 19.6℃, bp. 202℃, d_4^{20} 1.028, n_D^{20} 1.5372.

【Notes】

（1）It is best to use benzene without thiophene in this experiment. To remove thiophene from benzene, it can be washed with concentrated sulfuric acid multiple times（each time using concentrated sulfuric acid equivalent to 15% of the volume of benzene）, until there is no thiophene left. Then wash it with water, 10% sodium hydroxide solution and water in sequence, dry it with anhydrous calcium chloride, and then distill.

Method for testing thiophene in benzene: Take 1 mL of the sample, add 2 mL of a 0.1% indigo carmine solution in concentrated sulfuric acid, shake for several minutes. If there is thiophene, the acid layer will show a light blue-green color.

（2）Anhydrous aluminum trichloride exposed to air is extremely prone to absorb moisture and decompose, leading to ineffectiveness. Freshly sublimated or tightly packaged reagents should be used. The action should be quick when weighing. Blocky anhydrous aluminum trichloride should be quickly ground in the mortar before weighing.

（3）To add anhydrous aluminum trichloride, a simple feeder can be made: take a glass tube about 6 cm long and about 15 mm in diameter（which can be inserted into the side opening of the flask), and fit it with rubber stoppers at both ends. When weighing the drug, place it in the glass tube and tighten the stoppers at both ends of the glass tube. When adding the drug, open one stopper, quickly insert the glass tube into the bottle, gently tap the glass tube to make the drug enter the flask without sticking to the bottle's mouth. If there is solid residue in the tube that does not fall, open the other stopper and use a glass rod to push the solid down.

(4) Instruments or reagents that are not dried will seriously affect the experimental results or make the reaction difficult to carry out.

(5) This experiment can also be replaced with manual shaking instead of mechanical stirring. Use a 100 mL round-bottom flask with a two-neck connecting tube, with a dropper funnel on the positive neck and a reflux condenser on the side neck, and a calcium chloride drying tube and gas absorption device connected to the upper port of the condenser. In order to facilitate the oscillation of the reaction substance, the flask, condenser, and dropper funnel should be installed on the same iron frame. When using manual shaking, the reflux time should be extended.

(6) The yield can also be increased by prolonging the reflux time.

【Thought questions】

(1) Why is excess benzene and anhydrous aluminium chloride used?

(2) Why is acetic anhydride added slowly?

4.6 Preparation of adipic acid

【Experimental objectives】

To learn the preparation of acid by means of oxidation and the basic operation of recrystallization.

【Experimental principle】

The most commonly used method for preparing carboxylic acids is oxidation of alkenes, alcohols, and aldehydes. Common oxidizing agents include nitric acid, sulfuric acid solutions of potassium (or sodium) dichromate, potassium permanganate, hydrogen peroxide, and peracetic acid, among others.

Adipic acid is one of the main raw materials for the synthesis of nylon-66 and polyurethane. In the laboratory, it can be prepared by oxidizing cyclohexanol with nitric acid or potassium permanganate. When using nitric acid, toxic nitrogen dioxide gas is produced, so ventilation or exhaust gas absorption should be taken into consideration.

$$\text{C}_6\text{H}_{11}\text{–OH} \xrightarrow{[O]} \text{C}_6\text{H}_{10}\text{=O} \xrightarrow{[O]} HOOC(CH_2)_4COOH$$

【Chemicals】

Cyclohexanol	2.1 mL (2 g, 0.02 mol)
Nitric acid (d=1.42)	5 mL (0.08 mol)

【Time】 2 h

【**Experimental procedure**】

The reaction must be carried out inside the ventilating cabinet. The reaction conditions must be strictly obeyed.

In a 50 mL RBF, a thermometer is placed inside. The mercury ball is placed as close to the bottom of the rbf as possible.

Add 5 mL water and 5 mL nitric acid into the rbf and mix together. Heat the mixture to 80℃, add 2 drops of cyclohexanol inside with a dropper . The reaction set in immediately. The temperature rises to 85～90 ℃. 2.1 mL cyclohexanol is added carefully so that the temperature can stay within the above range. Add cold water if necessary. When alcohol is added completely and the temperature of the solution decreases below 80℃, heat the mixture to 85～90℃ for another 2～3 minutes.

Cool the solution in an ice bath. Collect the precipitate by means of suction. The precipitate was washed with 3 mL ice water twice. Then dry.

Yield: 1.4 g.

Data: colorless crystal, mp 153℃.

4.7 Preparation of acetic *n*–butyl ester.

【**Experimental objectives**】

To learn the esterification reaction.

【**Experimental principle**】

Carboxylic acid esters are a wide range of organic compounds. Many esters are used as raw materials for flavoring agents in food or cosmetics. Oils and fats are not only used as edible oils but also important industrial raw materials. Di-*n*-butyl phthalate and di-octyl phthalate are plasticizers for artificial leather and polyvinyl chloride（PVC）. Pyrethroids and their analogs are a type of highly effective, low-toxicity, and low-residue pesticides.

Carboxylic acid esters are commonly prepared by direct esterification of carboxylic acids and alcohols under acid-catalyzed conditions. Common catalysts include sulfuric acid, anhydrous hydrogen chloride, para-toluenesulfonic acid, and strongly acidic resins.

This reaction is reversible, and if equimolar amounts of carboxylic acid and alcohol are used, only about 2/3 of the starting materials will be converted to esters at equilibrium. To improve yield, one reactant is usually present in excess（based on factors such as availability, cost, and ease of separation from the product）. Removing the water generated during the reaction is also a common method to increase the yield. In certain esterification reactions, binary or ternary azeotropes can form with alcohol, ester, and water, or a solvent（such as benzene）that forms azeotropes with water can be added to the reaction system. Using a Dean-Stark trap can achieve satisfactory results.

Experimental evidence has shown that the excess addition of the catalyst in the reaction system can increase the reaction rate. The hydration of acid with water can remove some of the water generated in the reaction, thereby increasing the yield of esters.

【Reaction】

$$CH_3COOH + n\text{-}C_4H_9OH \rightleftharpoons CH_3COOC_4H_9\text{-}n + H_2O$$

【Chemicals】

n-butanol 11.5 mL（9.3 g, 0.125 mol）

Glacial acetic acid 7.2 mL（7.5 g, 0.125 mol）

Concentrated sulfuric acid, 10% sodium carbonate, anhydrous magnesium sulfate.

【Time】4 h

【Experimental procedure】

In a dry 50 mL RBF, add 11.5 mL *n*-butanol and 7.2 mL glacial acetic acid, 3～4 drops of concentrated sulfuric acid is added subsequently. Zeolite is added. Install the water knockout vessel and refluxing condenser. Add water in the knockout vessel to just below the outlet of the side tube. Heat the mixture to reflux, release the water inside the knockout vessel after a while to keep the surface level. No water will be produced in 40 min., which shows the reaction is complete. Stop heating, record the volume of the water separated. Let the mixture cool down to room tempreture. Pour the mixture in the water knockout vessel and the reactant in the rbf into a separatory funnel, wash with 10 mL water, get rid of the aqueous layer. The organic layer is washed with 10 mL of 10% sodium carbonate to check whether it is still acidic. Get rid of the aqueous layer, wash the organic layer with 10 mL water once again. The aqueous layer is separated. Pour the organic layer into a conical flask, anhydrous magnesium sulfate is added.

Pour the dry ester into a dry 50 mL RBF, zeolites is added, install the distillation device and start to distill, collect thes parts of 124～126℃.

Yield: 10～11 g.

Data: Colorless liquid, bp 126.5℃, d_4^{20} 0.882, n_D^{20} 1.3951.

【Note】

This experiment utilizes an azeotropic mixture to remove water from the esterification reaction, the lower layer is water.

Azeotropic mixture		Bp/℃	Composition（%）		
			Acetic *n*-butylester	*n*-butanol	water
Binary	Acetic *n*-butylester/water	90.7	72.9		27.1
	n-butanol/water	93.0		55.5	44.5
	acetic *n*-butyl ester/n-butanol	117.6	32.8	67.2	
Ternary	acetic *n*-butyl ester /*n*-butanol/water	90.7	63.0	34.6	29.0

4.8 Preparation of Ethyl acetoacetate

【Experimental objectives】

To learn the operations of sodium and review the operations of low pressure distillation.

【Experimental principle】

Esters containing α-active hydrogen can undergo Claisen ester condensation reactions with another molecule of ester in the presence of strong alkaline reagents such as $NaNH_2$, NaH, $NaOCH_2CH_3$, or triphenylmethyl sodium, resulting in the formation of *β*-keto esters. Ethyl 3-oxobutanoate is prepared through this reaction.

Typically, ester and metallic sodium are used as starting materials, and excess ester is used as a solvent. The reaction utilizes the trace amount of alcohol present in the ester to react with metallic sodium, forming sodium alkoxide. As the reaction progresses, the continuous generation of alcohol allows the reaction to proceed until the metallic sodium is consumed.

However, if the ester used as the raw material contains a high amount of alcohol, it will affect the yield of the product. Therefore, it is generally required that the alcohol content be below 3%.

The obtained ethyl 3-oxobutanoate is a mixture of keto and enol forms, with 93% of the keto form and 7% of the enol form at room temperature.

$$2CH_3COOC_2H_5 + C_2H_5ONa \longrightarrow CH_3C(ONa){=}CHCOOC_2H_5 + 2C_2H_5OH$$

$$CH_3C(ONa){=}CHCOOC_2H_5 + CH_3COOH \longrightarrow CH_3C(=O)CHCOOC_2H_5 + CH_3COONa$$

【Chemicals】

Acetic ethyl ester	48.9 mL（44 g, 0.5 mol）
Sodium	5 g（0.22 mol）

Dilute acetic acid, 5% sodium carbonate, anhydrous potassium carbonate, saturated sodium chloride, saturated calcium chloride.

【Time】 8 h

【Experimental procedure】

All chemicals used must be anhydrous, all the apparatuses are dry.

In a dry 250 mL RBF, add 48.9 mL anhydrous acetic ethyl ester and 5 g chopped sodium. Install the refluxing condenser immediately, place a dry tube with calcium chloride on the top of the condenser. According to scheme 2-4. Heat the mixture to start the reaction. If the reaction is too vigorous, remove the heating device and cool down the rbf with an ice bath. Keep the mixture refluxing gently. As soon as the sodium disappears, stop heating. The mixture turns red and transparent（sometimes yellow precipitates appeared）. Let the mixture cool down to room temperature and place the rbf into an ice bath, add diluted acetic acid slowly while shaking. When the mixture becomes weak acidic, all the solid inside dissolves, separate the red ester layer with a separatory funnel. Extract the ester layer into the aqueous layer with acetic ethyl ester. Combine the two parts of the ester layer. Wash the ester layer with 5% sodium carbonate to become neutral. Dry it with anhydrous potassium carbonate or magnesium sulfate.

Pour the dry liquid into a 125 mL distillation flask, install a low-pressure distillation device（Section2.5）. Distile out ethyl acetate at normal pressure. Then distil out ethyl acetoacetate at low pressure. The boiling point range of the collected fraction depends on the pressure shown as follows:

Pressure/kPa	1.666	1.866	2.399	3.866	5.998	10.66
Pressure/mmHg	12.5	14	18	29	45	80
bp/°C	71	74	79	88	94	100
bp range/°C	69～73	72～76	77～81	86～90	92～96	98～102

Yield: 8～9.5 g.

Pure ethyl acetoacetate is a kind of colorless liquid, bp 180°C/0.1006MPa（755 mmHg）（slightly decompose upon distillation）, d_4^{20} 1.025, n_D^{20} 1.4194.

【Notes】

（1）The ethyl acetate used must be anhydrous. The purification method is as follows: Wash the regular ethyl acetate several times with a saturated calcium chloride solution, then dry it with fused anhydrous potassium carbonate, and distill it in a water bath, collecting fractions at 76～77°C.

（2）The size of the sodium metal particles directly affects the rate of condensation reaction. Use forceps to remove the sodium metal chunks from the bottle, absorb any solvent oil with a double-layer filter paper, and then use a knife to remove the surface oxide layer of the sodium metal. Weigh it quickly and immediately press it into the flask using a sodium metal presser（or press it into anhydrous ether）. If there is no sodium metal presser, the sodium metal can be cut into fine strips（in cold anhydrous ether）and immediately transferred into a flask containing anhydrous ethyl acetate（try to minimize the exposure time to air）.

（3）The time required for complete disappearance of sodium metal depends on the size of the sodium particles, generally ranging from 1.5 h to 3 h.

（4）The condensation reaction step must be completed within one laboratory session, otherwise it will affect the yield.

（5）When adding acetic acid dropwise, extra caution is needed. If there is a small amount of unreacted sodium metal present in the reactants, a vigorous reaction may occur. Excessive addition of acetic acid should also be avoided, as it increases the solubility of the ester in water, resulting in a decreased yield. 30% acetic acid can be used at high room temperatures, while 25% acetic acid can be used at low room temperatures.

（6）The phenomenon of tautomeric isomerism in ethyl acetoacetate can be observed through a simple experiment as follows: Dissolve 2～3 drops of ethyl acetoacetate in 2 mL of water, add 1 drop of 1% ferric chloride solution, and observe the color change of the solution. Then quickly add bromine water until the color of the solution fades away. Allow it to stand still and observe the color change. Once the color reappears, bromine water can be added again, repeating the experiment multiple times.

【Thought questions】

（1）What's the effect if the apparatuses are not dry?

（2）Why is the low-pressure distillation used?

4.9 Preparation of cyclohexanone

【Experimental objectives】

To learn the operations of steam distillation.

【Experimental principle】

Cyclohexanone belongs to the class of cyclic ketones, having a similar odor to acetone. It is a colorless or pale yellow, transparent oily liquid at room temperature, slightly soluble in water and soluble in various organic solvents. Cyclohexanone is used as a raw material for the preparation of adipic acid and caprolactam.

In the laboratory, corresponding aldehydes or ketones are commonly prepared by oxidizing primary alcohols or secondary alcohols. Cyclohexanol is a secondary alcohol that can be oxidized to cyclohexanone by acidic sodium chromate (or potassium) . Ketones have relatively high stability and are generally not easily further oxidized. However, it is crucial to strictly control the reaction conditions to prevent excessive oxidation, as it may lead to the further oxidation and carbon chain cleavage of the product.

【Reaction】

$$\text{C}_6\text{H}_{11}\text{-OH} \xrightarrow{Na_2Cr_2O_7/H_2SO_4} \text{C}_6\text{H}_{10}\text{=O}$$

【Chemicals】

Cyclohexanol	10.4 mL（10 g, 0.1 mol）
Sodium bichromate（$Na_2Cr_2O_7.2H_2O$）	10.4 g（0.035 mol）
Concentrated sulfuric acid（*d*=1.84）	10 mL

Methanol, NaCl, anhydrous magnesium sulfate.

【Time】 6 h

【Experimental procedure】

In a 250 mL RBF, add 60 mL of ice water, add 10 mL concentrated sulfuric acid by dropwise slowly while shaking. 10.4 mL cyclohexanol is added slowly and carefully afterwards. Cool down the solution to 15°C.

In a 100 mL beaker, dissolve 10.4 g sodium bichromate in 10 mL water. Cool down the solution to 15°C. Add the above solution to the mixture of cyclohexanol in sulfuric acid while shaking constantly to make the reaction mix completely. The addition process has been accomplished several times. When the first part is added, the temperature will rise automatically and the mixture turns dark green from orange. When the temperature reaches 55°C, cool down the mixture with an ice bath. Keep the temperature at 55～60°C. When the orange color of the solution disappears completely, add the second part of the oxidant reagent. Keep shaking the flask until the temperature decreases and continue the process for 5～10 minutes. Finally, add 1～2 mL methanol to reduce the excess oxidant reagent.

Add 50 mL water and zeolite, install a distillation device and start distilling. Collect the mixture of cyclohexanone and water. Add sodium chloride into the product to dissolve completely. Transfer the liquid into a separatory funnel and let it stand still. Collect the organic layer and dry it with anhydrous magnesium sulfate.

Distill, collect a fraction of 151～156℃.

Yield: 6 g.

Data: colorless liquid, bp 155.7℃, d_4^{20} 0.948, n_D^{20} 1.4507.

【Note】

(1) The reactants should not be excessively cooled to avoid the accumulation of unreacted chromic acid. When chromic acid reaches a certain concentration, the oxidation reaction can become very vigorous, posing a risk of losing control.

(2) Alternatively, 0.5～1 g of oxalic acid can be added.

(3) This distillation operation is essentially a simplified version of steam distillation. Cyclohexanone and water form an azeotropic mixture with a boiling point of 95℃, containing 38.4% cyclohexanone.

(4) The solubility of cyclohexanone in water is 2.4 g/100 g water at 31℃. Adding table salt to the distillate reduces the solubility of cyclohexanone and facilitates its separation into distinct layers.

【Thought questions】

What's the purpose of adding methanol when the oxidation reaction completes?

第五部分

有机化学系列实验

5.1 系列实验一

5.1.1 溴乙烷的制备

【实验目的】

通过卤烷的制备掌握醇与氢卤酸发生的亲核取代反应。

【实验原理】

醇和氢溴酸作用可以生成溴代烷，实验室可以用浓氢溴酸（质量分数 47.5%）稀释，或通过溴化钠与硫酸反应制备氢溴酸。

反应式：

$$NaBr+H_2SO_4 \longrightarrow HBr+NaHSO_4$$

$$C_2H_5OH+HBr \rightleftharpoons C_2H_5Br+H_2O$$

虽然上式反应是可逆的，但是，可以采用增加其中一种反应物的浓度或设法使产物溴乙烷及时离开反应体系的方法，使平衡向右移动。本实验正是这两种措施并用，以使反应顺利完成。

此外，还存在下列副反应：

副反应：

$$2C_2H_5OH \xrightarrow{H_2SO_4} C_2H_5—O—C_2H_5+H_2O$$

$$C_2H_5OH \xrightarrow{H_2SO_4} C_2H_4+H_2O$$

【药品】

乙醇	7.9 g（10 mL，0.165 mol）
无水溴化钠	13 g（0.126 mol）
浓硫酸	19 mL
亚硫酸氢钠	

【实验步骤】

用一锥形瓶，加入 9 mL 水。在冷却和振荡下，慢慢加入 19 mL 硫酸。采用 250 mL 圆底烧瓶来搭设蒸馏装置。将 13 g 研细的溴化钠倒入烧瓶中，再将 10 mL 乙醇加入，最后倒入冷却好的硫酸溶液，振荡，以防止溴化钠结块。要紧密装好蒸馏装置。为防止溴乙烷挥发，在接收器内加入冷水后，置于冰水浴中，并使接引管的末端刚浸没在接收器的水面下。

在电热套中低温加热，缓慢使其发泡。当气泡渐多时，调节电热套的温度，慢慢蒸出反应的油状物。当反应不太激烈时可适当提高电热套温度，使其反应完全，直到无油滴蒸出（可用干净的表面皿放入一些净水，从接引管末端接几滴馏出液，如无油状物即为反应完全）。

反应完全后，先将馏出液移去，再停止加热。馏出液用分液漏斗将其分离。下层的粗溴乙烷放入干燥的锥形瓶中。在冰水冷却下逐滴加入浓硫酸，同时振荡，直到溴乙烷变得澄清且有液层分出时为止（约需 3～4 mL 硫酸）。用干燥的分液漏斗小心地将硫酸分出。从上口将溴乙烷倒入干燥的 50 mL 圆底烧瓶内，在干燥的蒸馏装置中用水浴加热，蒸出溴乙烷。收集 37～40℃馏分。

溴乙烷为无色液体，沸点 38.4℃，d_4^{20} 1.46，n_D^{20} 1.4239。

【测试与检验】

（1）铜的卤化物生成——鄂尔斯坦试验

取一段细铜丝，将其一端绕成 2～3 圈螺旋，在火焰上加热至火焰无色。冷却后浸入溴乙烷中，然后将其放入火焰中。观察火焰的颜色变化。

（2）硝酸银酒精溶液试验

卤代烃及其衍生物可与硝酸银作用生成卤化银沉淀：

$$RX + AgNO_3 \longrightarrow AgX\downarrow + RONO_2$$

取 0.5 mL 硝酸银酒精溶液，放入洗净干燥的试管中，加入 2 滴新制的溴乙烷样品，振荡后，再静置约 3 min，观察有无沉淀。如不析出沉淀，在水浴中温热 2～3 min，再观察现象。

（3）碘化钠的丙酮溶液试验

氯化物及溴化物能与碘化钠反应生成氯化钠和溴化钠：

$$RCl + NaI \xrightarrow{\text{丙酮}} RI + NaCl\downarrow$$

$$RBr + NaI \xrightarrow{\text{丙酮}} RI + NaBr\downarrow$$

取 0.5 mL 碘化钠丙酮溶液放入干燥试管中，加入 2 滴溴乙烷，振荡后，静置约 2 min，观察有无沉淀，在水浴上加热，观察现象（如有淡红色出现，表明有碘析出）。

【注意事项】

（1）本实验及“苯乙醚的制备”实验既可单独开设，也可以组成一个系列实验。

（2）如用含有结晶水的 NaBr，应将含水量扣除。

（3）如果油状物为棕黄色，说明溴被分解出来。可以加入亚硫酸氢钠除去颜色。

（4）如发生倒吸，可将接引管轻轻移动，使倒吸上来的液体回到接收器中，再适当调大火焰，使其恢复正常反应。

【思考题】

（1）制备溴乙烷时为什么加入一定量的水？

（2）粗溴乙烷中加硫酸的目的是什么？

（3）本实验极易产率不高，分析其原因。

5.1.2 苯乙醚的制备

【实验目的】

掌握威廉森合成醚的方法，熟悉使用电磁搅拌的方法。

【实验原理】

醚可以看做是两分子醇之间失去一分子水后生成的化合物，也可以看做羟基化合物（醇、酚、萘酚等）中羟基氢被烃基取代的衍生物。若醚中的两个基团相同，则该醚称为单醚或对称醚；若两个基团不同，则称为混合醚或不对称醚。单醚通常采用在酸催化下醇分子间脱水来制备；混合醚则可采用醇盐和卤代烷的反应来制备。

苯酚与氢氧化钠形成酚钠，酚钠与溴乙烷经威廉森合成法生成混醚。

$$C_6H_5OH + NaOH \longrightarrow C_6H_5ONa + H_2O$$

$$C_6H_5ONa + C_2H_5Br \longrightarrow C_6H_5OC_2H_5 + NaBr$$

【药品】

苯酚	7.5 g（0.08 mol）
氢氧化钠	4 g（0.10 mol）
溴乙烷	12.4 g（8.5 mL，0.12 mol）
饱和氯化钠水溶液	
无水氯化钙	

【实验步骤】

按图 2-5 安装实验装置。先将磁力搅拌子放入 250 mL 三口烧瓶内，然后加入 7.5 g 苯酚、4 g 氢氧化钠和 4 g 水。装好恒压滴液漏斗和回流冷凝器，恒压滴液漏斗中加入 8.5 mL 溴乙烷。

在水浴中加热，使固体全部溶解，然后开启磁力搅拌器进行搅拌。调节水浴温度，使之保持在 80～90℃之间，然后缓慢滴加溴乙烷，约用 1 h。滴加完毕后，继续保持温度搅拌 1.5 h。液体冷却后，加约 10 mL 水使析出的固体完全溶解。将液体倒入分液漏斗，分出水层。有机层用等体积的饱和氯化钠水洗涤两次。分出粗产物，用无水氯化钙干燥。

将干燥好的粗产物进行蒸馏，收集 165～173℃馏分。如进行减压蒸馏，可参考表 1 苯乙醇压力与沸点的关系进行蒸馏。

表 1 苯乙醇的压力与沸点关系表

压力/kPa	0.67	1.33	2.67	5.33	8.00	13.33	26.6
沸点/℃	43.7	56.4	70.3	86.6	95.4	103.4	127.9

纯苯乙醚为无色液体，沸点 170℃，d_4^{20} 0.966，n_D^{20} 1.5076。

【思考题】

反应过程中，回流的液体是什么？产生的固体是什么？为什么反应后期回流不太明显？

5.2 系列实验二

5.2.1 溴苯的制取

【实验目的】

掌握芳烃卤代反应。

【实验原理】

芳香族卤代物的制法和卤代烷不同，一般是用卤素（氯或溴）在铁粉或三卤化铁催化下与芳香族化合物作用，通过芳香烃的亲电取代反应将卤原子直接引入芳环。真正的催化剂是三卤化铁。铁粉先和卤素作用生成三卤化铁，然后三卤化铁再起催化作用，整个取代反应的历程是：

$$2Fe+3Br_2 \longrightarrow 2FeBr_3$$

$$FeBr_3+Br_2 \rightleftharpoons Br^+[FeBr_4]^-$$

$$C_6H_6 + Br^+ \rightleftharpoons [C_6H_6Br]^+ \longrightarrow C_6H_5Br + H^+$$

$$[FeBr_4]^-+H^+ \longrightarrow FeBr_3+HBr$$

由于三卤化铁和卤素作用生成卤素正离子和四卤化铁复合负离子要一定的时间，因此在卤代反应开始前有一个诱导期。例如在制溴苯时，开始时反应不明显，过一段时间后反应进行很剧烈。反应过程中，需要将溴慢慢滴加到过量的苯中来避免反应过于剧烈和减少副产物二溴代苯的生成。三卤化铁很容易水解，故所用的试剂和仪器都必须是无水和干燥的。

$$C_6H_6 + Br_2 \longrightarrow C_6H_5Br + HBr$$

副反应：

$$2C_6H_5Br + Br_2 \longrightarrow p\text{-}C_6H_4Br_2 + o\text{-}C_6H_4Br_2 + 2HBr$$

【药品】

苯（无水）	10 g（11.5 mL，0.13 moL）
溴	16 g（5 mL，0.1 mol）
铁粉	0.25 g
饱和亚硫酸氢钠溶液	
无水氯化钙	

【实验步骤】

用干燥的 250 mL 三口烧瓶，安装搅拌器、回流冷凝管和恒压滴液漏斗。回流冷凝管上口连接一弯管引入到盛有水的锥形瓶中，用来吸收溴化氢气体，但管口不要插到液面里。

在三口烧瓶内加入 11.5 mL 无水苯和 0.25 g 铁屑，恒压滴液漏斗中加入 5 mL 溴。

在三口烧瓶内先滴加约 1 mL 溴，随即反应开始（必要时用水浴温热，约 30～40℃），可观察到有溴化氢气体冒出。在搅拌下，慢慢滴入剩下的溴，使溶液保持微沸（约 30 min）。加完溴后，再在水浴中（约 60～70℃）加热，直到液面不再有红色溴蒸气及溴化氢气体冒出为止，此时反应完成。将反应物冷却到室温，然后在三口烧瓶中加入 15 mL 水，搅拌约 2 min，将粗产物倒入分液漏斗，分出水层，粗溴苯用 10 mL 饱和亚硫酸氢钠溶液洗涤两次，使油层呈浅黄色为止。用 10 mL 水洗涤，分出溴苯并用无水氯化钙干燥。

将干燥好的粗溴苯在电热套中进行蒸馏（用空气冷凝器），收集 140～170℃馏分，再将 140～170℃馏分重新蒸馏一次，收集 154～160℃馏分。

纯溴苯为无色油状液体，沸点 150℃，d_4^{20} 1.499，n_D^{20} 1.5597。

【注意事项】

（1）本实验与“三苯甲醇的制备”实验可以组成为一个系列实验。

（2）量取溴必须在通风橱内进行，要戴好防护镜及手套，注意不要吸入溴蒸气。如接触到溴液可用硫代硫酸钠溶液洗涤。

（3）用水洗涤可以除去溴化铁，同时可以除去溴化氢和部分溴。

（4）用饱和亚硫酸氢钠洗涤可以除去溴。因溴在溴苯中溶解度比在水中大，所以很难用水把溴洗干净。

【思考题】

（1）制备溴苯实验中，哪个试剂是过量的？

（2）为什么要用干燥的烧瓶，苯和溴中如含有水分，对实验有何影响？

5.2.2 三苯甲醇的制备

【实验目的】

学习并掌握格利雅反应。

【实验原理】

卤代烷在无水四氢呋喃或者无水乙醚中和金属镁作用，生成烷基卤化镁 RMgX（Grignard 试剂）。用 Grignard 试剂与醛酮或环氧烷反应是制备各种复杂结构醇的常用方法。因为水、氧或其他活泼氢的存在都会破坏 Grignard 试剂，反应必须在无水、无氧、无活泼氢的条件下进行。在惰性气体（如氩气、氮气）下进行反应，可有效防止 Grignard 试剂与氧和二氧化碳的反应。偶合反应不可避免，可用稀溶液来减少这种副反应的发生。所以在反应真正引发以前，不能加过多的卤代烃。烷基碘比烷基溴、烷基氯更易发生偶合反应，在制备 Grignard 试剂中优先采用活性较低的烷基溴和烷基氯。

Grignard 试剂与碳基化合物形成的加成物，在酸性条件下进行水解，由于水解时放热，故要在冷却下进行，对遇酸极易脱水的醇，最好用饱和氯化铵溶液进行水解。

Grignard 试剂与羧酸酯等进行加成反应，再将加成物水解，即可得到叔醇。

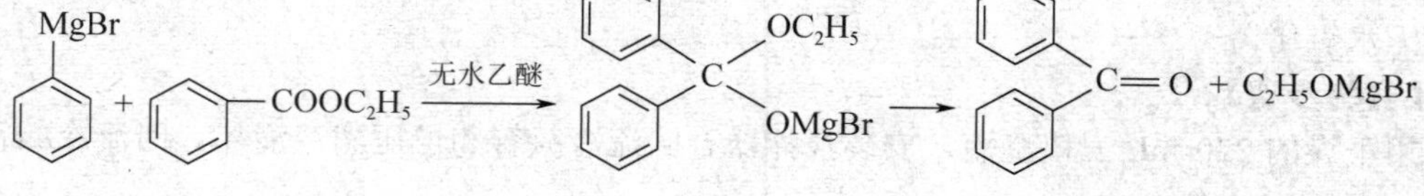

$$C_6H_5MgBr + C_6H_5COOC_2H_5 \xrightarrow{\text{无水乙醚}} (C_6H_5)_2C(OC_2H_5)(OMgBr) \longrightarrow (C_6H_5)_2C{=}O + C_2H_5OMgBr$$

$$(C_6H_5)_2C{=}O + C_6H_5MgBr \xrightarrow{\text{无水乙醚}} (C_6H_5)_2C(OC_2H_5)(OMgBr) \xrightarrow{NH_4Cl，H_2O} (C_6H_5)_3C{-}OH$$

【药品】

镁屑	1.5 g
溴苯	10.49 g（7 mL，0.068 mol）
无水乙醚	21.39 g（30 mL，0.289 mol）
苯甲酸乙酯	4.20 g（4 mL，0.016 mol）
氯化铵	7.5 g
碘少量	

【实验步骤】

按图 2-6 和图 2-7 安装好搅拌反应装置，回流冷凝器上口装一干燥管和吸收装置。将 1.5 g 镁屑，1～2 粒碘放置于 250 mL 三口烧瓶中。往三口烧瓶一端的恒压滴液漏斗中加入 7 mL 溴苯及 25 mL 无水乙醚混合液，先滴入 5 mL 混合液于三口烧瓶中，片刻后即发生反应，碘的颜色逐渐消失。如不发生反应，可小火加热水浴促使其反应。当反应开始后，开动搅拌装置，并将剩余溴苯乙醚混合液逐滴滴入，保持反应液处于微沸状态。加完后，控制加热温度，水浴加热回流半小时，至镁屑作用完全为止。

将三口烧瓶用冰水浴冷却，在滴液漏斗中加入 4 mL 苯甲酸乙酯及 5 mL 无水乙醚混合液。在搅拌下，逐滴加入三口烧瓶中。加完后，加热回流 0.5 h。

将三口烧瓶用冰水浴冷却，在滴液漏斗中加入新配制的氯化铵饱和溶液 30～35 mL，分解加成产物。将反应混合物倒入分液漏斗，分出醚层，倒入 100 mL 圆底烧瓶中，进行水浴蒸馏，蒸去乙醚，回收。最后，进行水蒸气蒸馏，除去未反应的溴苯和副产物联苯，反应至无油珠状物质馏出为止。此时，反应瓶中三苯甲醇呈固体析出，待冷却后，减压过滤并用水洗涤固体 2～3 次。粗产物用乙醇-水重结晶得到白色棱状三苯甲醇结晶。

纯三苯甲醇的熔点为 164.2℃。

【注意事项】

（1）本实验必须无水操作，各反应仪器与试剂须充分干燥后使用。

（2）镁屑采用新制的。如表面有氧化镁层，可用 5%盐酸溶液作用数分钟，去除酸液后，依次用水、乙醇、乙醚洗涤，吹干即可。

（3）由于溴苯的 Grignand 反应不易发生，故加入少量碘粒引发反应。

（4）为防止反应过于剧烈而增加副产物的生成，溴苯乙醚混合液滴加不宜过快。

（5）为防止氯化铵饱和溶液分解失去作用，应随配随用，取 7.5 g 氯化铵溶于 32 mL 水中。

（6）因为过程为放热反应，应慢慢加入氯化铵溶液，否则由于反应剧烈而使乙醚冲出。如反应中仍有絮状氢氧化镁未溶，可加入少许稀盐酸溶液，至全部溶解为止。

【思考题】

（1）制备 Grignand 试剂时，溴苯滴入太快或一次性加入是否可以？其后果是什么？

（2）本实验中，在制备 Grignand 试剂时同时有哪些副反应发生？如何避免？

（3）本实验采用饱和氯化铵分解加成产物，还可以采用什么试剂？

（4）本实验水解前各步，要求仪器、试剂必须干燥的原因是什么？应采取什么方法？
（5）在制备苯基溴化镁时，采取了什么措施来引发反应，还有什么方法？
（6）在用混合溶剂进行重结晶时，何时加入活性炭为宜？采用何种溶剂洗涤结晶？

5.3 系列实验三

5.3.1 硝基苯的制备

【实验目的】

掌握苯的硝化反应，熟悉空气冷凝器的使用。

【实验原理】

芳香族硝基化合物一般是由芳香族化合物直接硝化制备。根据被硝化物的活性，采用不同的硝化剂。浓硝酸与浓硫酸的混合液，是常用的硝化试剂之一，亦称混酸。硝化反应是不可逆反应。混酸中浓硫酸不仅可以用于脱水，而且有利于 NO_2^+ 的生成，从而提高反应速率。

硝基苯是最简单易得的芳香硝基化合物。无论是实验室制备还是工业生产，都是用苯与混合酸作用，在 50℃左右进行硝化。若反应时温度过高，易生成间二硝基苯。

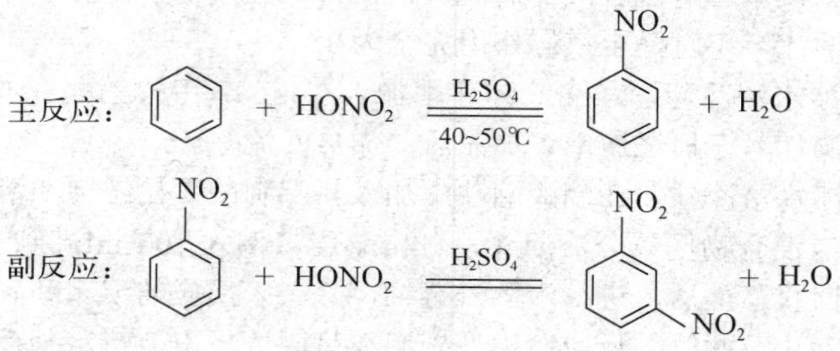

主反应：$C_6H_6 + HONO_2 \xrightarrow[40\sim50℃]{H_2SO_4} C_6H_5NO_2 + H_2O$

副反应：$C_6H_5NO_2 + HONO_2 \xrightarrow{H_2SO_4} m\text{-}C_6H_4(NO_2)_2 + H_2O$

【药品】

苯	23.7 g（27 mL，0.28 mol）
硝酸	30 mL（*d*=1.40，0.67 mol）
浓硫酸	38 mL（*d*=1.84，0.71 mol）
碳酸钠溶液（10%）	
饱和氯化钠溶液	
无水氯化钙	

【实验步骤】

首先配制混酸溶液。取 38 mL 浓硫酸，分批（每次约 4 mL）倒入盛有 30 mL 浓硝酸的锥形瓶中，将锥形瓶用冷水浴冷却，并不断地振荡。静置冷却。

按图 2-6 装置安装各仪器。在 250 mL 三口烧瓶中，加入 27 mL 苯，打开搅拌装置，将新配制的混酸倒入恒压滴液漏斗中，并通过恒压滴液漏斗缓慢加入三口烧瓶中。为使反应物温度控制在 40～50℃之间，用冷水浴冷却三口烧瓶。水浴加热到 60℃左右，反应半小时。

待反应物冷却后，用分液漏斗分去酸层，并回收。分出的粗硝基苯用 80 mL 水分两次洗涤，然后用约 10 mL 10% Na_2CO_3 水溶液洗涤，直到洗涤液不呈黄色为止。最后用水洗至中性。将粗硝基苯置于干燥的磨口锥形瓶中，加入适量的无水氯化钙干燥。

将干燥后澄清透明的硝基苯倒入 50 mL 圆底烧瓶中，加入少量沸石，连好空气冷凝管，在电热套中加热蒸馏，收集 205～210℃的馏分。注意：圆底烧瓶中要保留 1～1.5 mL 残液，不可蒸干。

纯硝基苯为无色（淡黄色）液体，具有苦杏仁气味，沸点为 210.9℃，d_4^{20}=1.203，n_D^{20}=1.5562。

【注意事项】

（1）本实验与“苯胺的制备”、“乙酸苯胺的制备”可组成一个系列实验。

（2）因硝化反应为放热反应，温度超过 50℃后会有较多的二硝基苯生成，同时将有部分硝酸和苯被挥发。

（3）苯的硝化反应是放热反应。开始加入混酸时，由于硝化反应较快，因此每次加入量不可太多。随着混酸的加入和硝基苯的生成，反应混合物中苯量逐渐降低，硝化速度减慢。故在加入一半混酸后，每次加入量可酌量增加。

（4）吸取少许上层反应液，滴入饱和 NaCl 溶液中，当油球下沉时就表示硝化反应基本完成。

（5）工业浓硫酸常常含有少量汞盐等杂质，这些杂质具有催化作用使反应中产物含有微量的多硝基酚，如苦味酸及二硝基酚等，它们的碱溶液呈黄色。

（6）硝基苯中夹杂的硝酸若不洗净，最后蒸馏时会分解，并产生 NO_2 气体，也增加生成二硝基苯的可能性。

（7）因硝基苯有毒，所以在处理时须十分小心。如溅到皮肤上，可用少量酒精擦洗，再用肥皂及温水洗净。

（8）经洗涤后的粗硝基苯因含小水珠，呈浑浊状态，加入干燥剂后，可用水浴温热（30～50℃）并摇动，以加速干燥，然后冷却放置待蒸。

（9）最后蒸馏温度不得超过 210℃，绝不可蒸干，以防爆炸。

【思考题】

（1）若将新配制的混酸一次加入苯中，会有什么结果？

（2）实验中，控制反应温度在 40～50℃之间的原因是什么？若温度过高，其结果如何？

（3）粗产物依次用水、碱液、水洗涤的目的是什么？它们在反应条件上有什么不同？

（4）在进行蒸馏精制时，为什么烧瓶残液不可蒸干？

5.3.2　苯胺的制备

【实验目的】

掌握芳香族硝基化合物的还原反应，学习水蒸气蒸馏的方法。

【实验原理】

芳胺的制备，一般不能直接将氨基（—NH_2）导入芳环上，而是经过间接的方法来制备。在酸性介质中将芳香硝基化合物还原就是制取芳胺的一种重要方法。

$$ArNO_2+6\,[H] \rightarrow ArNH_2+2H_2O$$

常用的还原剂有：铁-盐酸、铁-乙酸、锡-盐酸、氯化亚锡-盐酸等。由硝基苯制备苯胺时，常用铁屑在含有少量酸性电解质的溶液中进行。

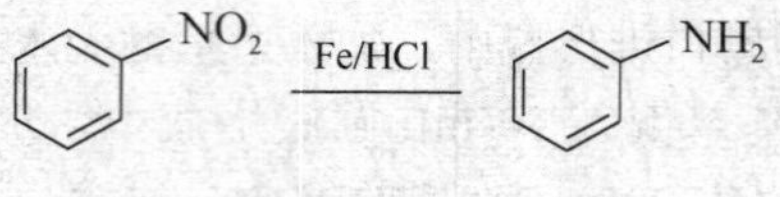

NO_2 $\xrightarrow{Fe/HCl}$ NH_2

【药品】

硝基苯	6.02 g（5 mL，0.05 mol）
细铁屑	15 g（0.27 mol）
浓盐酸	1 mL
氯化钠	
碳酸钠	
固体氢氧化钠	

【实验步骤】

按图 2-6 搅拌回流装置安装各仪器（注意：三口烧瓶一侧温度计的水银球应浸入液面下）。在三口烧瓶中，加入 25 mL 水、15 g 细铁屑、1 mL 浓盐酸及 5 mL 硝基苯，最后加入少量沸石。打开搅拌装置，在电热套中低温加热，用 5～10 min 将反应加热至沸。因本反应为放热反应，当开始回流时即撤去加热源。待反应趋于缓和时，加热 30 min 左右，反应混合物变为黑色，反应即基本完成。

待反应物冷却后，用 10 mL 水将回流冷凝器和搅拌棒上的黏着物冲洗到三口烧瓶中。取下三口烧瓶，一边摇动，一边加入约 1 g 碳酸钠粉末，直到混合物呈碱性为止。

将反应装置改装成水蒸气蒸馏装置（图 2-11），进行水蒸气蒸馏，待馏出液不再混浊为止，停止蒸馏。将食盐加到馏出液中，达到饱和后，静置。然后，用分液漏斗分出水层。将苯胺层倒入干燥的磨口烧瓶中，用适量固体氢氧化钠进行干燥。

将干燥好的粗苯胺倒入 50 mL 圆底烧瓶中加入少量沸石，电热套中加热蒸馏，收集 182～185℃的馏分。

纯苯胺为无色油状液体，沸点 184.4℃，d_4^{20}=1.022，n_D^{20} 1.5863。

【检验与测试】

（1）苯胺与溴水反应：5 mL 水中加入一滴苯胺，振荡使其溶解后，取 5 mL 苯胺水溶液，滴加饱和溴水溶液，立刻有白色的混浊物或沉淀析出。

（2）胺具有碱性，可与强酸生成水溶性的盐。由此可识别胺：5 mL 水加一滴苯胺，振荡使其溶解。加 6 滴浓盐酸，振荡，观察现象。最后用水稀释，溶液仍呈透明状。

【注意事项】

（1）本实验采用浓盐酸作催化剂，亦可用 1 mL 冰乙酸或 0.25～1 g 氯化铵代替。

（2）苯胺和硝基苯均有毒，操作时应小心，避免与皮肤接触或吸入其蒸气。若不慎触及皮肤时，应先用水冲洗，再用肥皂及温水洗净。

（3）此步骤主要是使铁活化，以缩短还原时间。

（4）当反应进行约 20 min 时，反应物转为褐色，以后颜色转深，最后变为黑色，反应完成。欲检验反应是否完成，可用吸管吸取少许反应混合物，滴入 1 $mol\cdot L^{-1}$ 盐酸中，当看不到油珠时，表明反应已完成。

（5）待反应物冷却后，再加入碳酸钠粉末，以防止泡沫溢出。

（6）反应完成后，三口烧瓶壁上将黏附黑褐色物质，可用 1∶1（体积比）的盐酸水溶液温热除去。

（7）在 22℃时，每 100 mL 水约溶解 3 mL 苯胺，为了减少苯胺的损失，根据盐析的原理，加入食盐，使馏出液饱和。每 100 mL 馏出液中加入研细的 NaCl 20～25 g。由此，原溶于水中的绝大部分苯胺即成油状物析出。

（8）由于苯胺的沸点较高，可采用空气冷凝器蒸馏。

【思考题】

（1）在反应过程中，为什么要充分搅拌反应混合物？

（2）在进行水蒸气蒸馏前，为什么采用碳酸钠处理反应物使之呈碱性？用什么试纸来检验？

（3）为什么用水蒸气蒸馏能把粗苯胺分离出来？

（4）若制得的苯胺中含有硝基苯，应如何除掉？

（5）蒸馏精制前，为什么用固体氢氧化钠进行干燥？

5.3.3　乙酰苯胺的制备

【实验目的】

掌握芳胺的乙酸化反应。

【实验原理】

有机合成中，芳胺的乙酰化常被用来“保护”氨基，以降低芳胺对氧化性试剂的敏感性，并适当降低芳环活性以有利于单取代产物的生成。同时由于乙酰基的空间效应，往往选择性地生成对位取代产物。反应完成以后，再水解除去乙酰基。

苯胺与冰醋酸的反应是可逆反应，为防止乙酰苯胺的水解，提高产率，本实验采用将反应物（醋酸）过量并将生成物（水）从反应中不断移出的方法，使平衡向右移动。由于水与醋酸（bp=117.9℃）的沸点相差不大，可利用分馏柱并严格控制分馏柱柱顶温度在 110℃以下来有效移出副产物水。

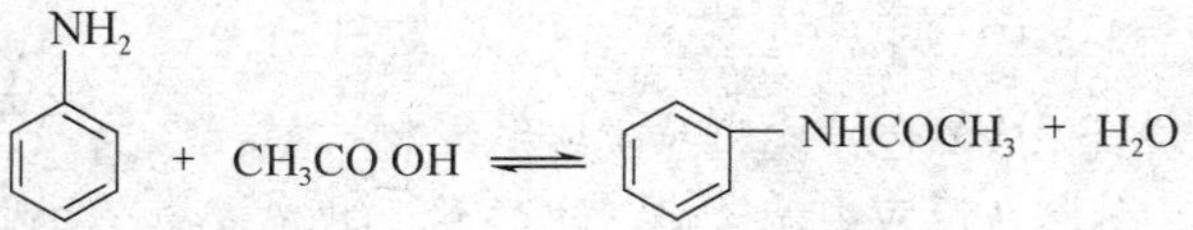

$$C_6H_5NH_2 + CH_3COOH \rightleftharpoons C_6H_5NHCOCH_3 + H_2O$$

【药品】

苯胺	5.1 g（5 mL，0.055 mol）
冰醋酸	8.3 g（8 mL，0.132 mol）
锌粉	
活性炭	

【实验步骤】

按图 2-10 在 50 mL 的圆底烧瓶上装好分馏柱、蒸馏头、温度计及接引管和接收瓶。

圆底烧瓶中放入 5 mL 新蒸过的苯胺和 8 mL 冰醋酸，再加入少量锌粉，电热套中加热至沸腾，调节电压保持温度在 105℃以上，不超过 110℃使反应生成的水充分蒸出来。约经 40～60 min，当温度计读数发生波动和减小，或容器内出现白雾状时说明反应已达到终点，停止加热。

在搅拌下将反应物缓慢地倒入盛有 100 mL 水的烧杯中，继续搅拌并冷却烧杯，使粗乙酰苯胺完全析出。用布氏漏斗抽滤，固体再用 5～10 mL 冷水洗涤，除去多余的酸。

将粗乙酰苯胺移入烧杯中，加入 80 mL 水，加热沸腾。如仍有油珠状物需补加热水，直到油珠在沸腾下全部溶解后再加入约 2 mL 水。稍冷，在搅拌下加入 0.5 g 活性炭，再煮沸约 5 min，趁热进行过滤。滤液冷至室温，乙酰苯胺呈片状晶体析出。减压过滤，尽量除去晶体中的水，产品放在表面皿上干燥。

纯乙酰苯胺为白色片状晶体，熔点 114℃。

【检验与测试】

（1）胺酰基化产物在酸性条件下水解，经亚硝酸钠处理形成重氮盐，再与β-萘酚反应，可形成鲜红色到深褐色的沉淀。

$$RCONHAr \xrightarrow{H^+} RCOOH + ArNH_3^+ \xrightarrow{NaNO_2} ArN_2^+ \xrightarrow{\beta-萘酚} \text{(1-(N=N-Ar)-2-萘酚)}$$

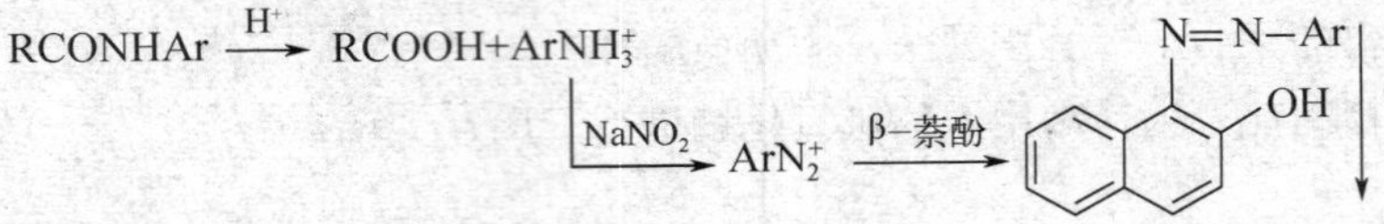

取一支试管，加入约 0.2 g 乙酰苯胺，再加入 1 mL 70%硫酸，加热使乙酰苯胺全部溶解。然后将溶液冷至 5℃，加 3～4 滴 5% $NaNO_2$ 水溶液，再加入几滴 5%的 *β*-萘酚的 10%NaOH 溶液。观察颜色变化。如有鲜红色到深褐色的沉淀说明乙酰苯胺已形成。

（2）红外光谱鉴别 作产物的红外光谱并与表 1 对比后鉴别。

表 1 乙酰苯胺的红外光谱中的振动频率

3200～3300 cm^{-1}	N—H 的伸缩振动
1670 cm^{-1}	C=O 伸缩振动
1560 cm^{-1}	N—H 的弯曲振动
760 cm^{-1}	苯取代的 C—H 弯曲振动
699 cm^{-1}	苯取代的 C—H 弯曲振动

【注意事项】

（1）久置的苯胺易氧化成深色，实验前应该先蒸馏一次。

（2）锌粉主要防止苯胺在反应中被氧化。注意不能加得太多，否则会形成氢氧化锌，不利于后处理。

（3）油珠为熔融状态的含水乙酰苯胺。乙酰苯胺在 100 mL 水中的溶解度为：25℃，0.563 g；80℃，3.5 g；100℃，5.2 g。

（4）为防止溶液过饱和，加入一定量的水有利于热过滤。

【思考题】

（1）本实验应注意什么才能使反应完全？

（2）重结晶操作中，应注意哪些事项才能使产率提高、质量好？

Serial organic experiments

5.1 Series Experiment Ⅰ

5.1.1 Preparation of bromoethane

【Experimental objectives】

To learn the nucleophilic substitution reaction between alcohols and hydrogen halide acid by means of the preparation of haloalkanes.

【Experimental principle】

Alcohols can react with hydrogen bromide to generate alkyl bromides. In the laboratory, concentrated hydrogen bromide（mass fraction 47.5%）can be used, or hydrogen bromide can be prepared by reacting sodium bromide with sulfuric acid.

Main reaction: $NaBr+H_2SO_4 \longrightarrow HBr+NaHSO_4$

$$C_2H_5OH+HBr \rightleftharpoons C_2H_5Br+H_2O$$

Although the above reaction is reversible, the equilibrium can be shifted to the right by increasing the concentration of one reactant or finding a way to remove the product ethyl bromide from the reaction system promptly. This experiment employs both of these measures to ensure the smooth completion of the reaction.

In addition, the following side reactions may also occur:

Side reaction: $2C_2H_5OH \xrightarrow{H_2SO_4} C_2H_5—O—C_2H_5+H_2O$

$$C_2H_5OH \xrightarrow{H_2SO_4} C_2H_4+H_2O$$

【Chemicals】

Ethanol	7.9 g（10 mL, 0.165 mol）
Anhydrous sodium bromide	13 g（0.126 mol）
Concentrated sulfuric acid	19 mL
Sodium bisulfite	

【Experimental procedure】

In a conical flask, 19 mL concentrated sulfuric acid is added into 9 mL water slowly and carefully while shaking and cooling. Install a distillation apparatus with a 250 mL RBF. 13 g powder sodium bromide, 10 mL ethanol, cold sulfuric solution are added separately and consequently into the RBF. Shake the mixture to avoid the agglomeration of sodium bromide. Seal the distillation apparatus. In order to avoid the evaporation of bromoethane, add cold water to the receptor and put the receptor in an ice bath. Place the terminal of the guide pipe just below the surface of water in the receptor.

Heat the mixture to bubble gently. Keep heating warmly to distill out the oily product.

When the reaction is complete, remove the distillation fraction first, then stop heating. Pour the distillation fraction into a separatory funnel and collect the organic layer. The crude product was treated by adding concentrated sulfuric acid（3～4 mL）and shaking in the meantime, until the organic layer becomes clear. Separate the organic layer with a dry separatory funnel. Pour the crude bromoethane into a dry 50 mL RBF, distill, collect a fraction of 37～40℃.

Data: colorless liquid, bp 38.4℃, d_4^{20} 1.46, n_D^{20} 1.4239.

【Testing and Analysis】

（1）Halide Formation Test for Copper——Erlenmeyer Test

Take a piece of fine copper wire and coil one end into 2～3 loops. Heat it in a flame until the flame becomes colorless. After cooling, immerse it in bromoethane and then place it back into the flame. Observe the color change of the flame。

（2）Silver Nitrate Ethanol Solution Test

Halogenated hydrocarbons and their derivatives can react with silver nitrate to form silver halide precipitates:

$$RX + AgNO_3 \longrightarrow AgX\downarrow + RONO_2$$

Take 0.5 mL of an ethanol solution of $AgNO_3$ and add it to a clean and dry test tube. Add 2 drops of freshly prepared bromoethane sample, shake it, and then let it stand for about 3 minutes to observe for any precipitation. If no precipitate forms, heat it in a water bath for 2～3 minutes and observe the phenomenon again.

（3）Sodium Iodide Acetone Solution Test

Chlorides and bromides can react with sodium iodide to generate sodium chloride and sodium bromide:

$$RCl + NaI \longrightarrow RI + NaCl\downarrow$$

$$RBr + NaI \longrightarrow RI + NaBr\downarrow$$

Take 0.5 mL of sodium iodide acetone solution and place it in a dry test tube. Add 2 drops of bromoethane, shake it, and let it stand for about 2 minutes to observe for any precipitation. Heat it in a water bath and observe the phenomenon（if a pale red color appears, it indicates the presence of iodine precipitate）.

【Notes】

（1）This experiment, along with the "Preparation of Phenyl Ethyl Ether" experiment, can be conducted separately or as a series of experiments.

（2）If using NaBr with crystalline water, the amount of water should be taken into account and deducted accordingly.

(3) If the oily substance appears brownish-yellow, it indicates that bromine has been released. You can add hydrogen sulfite solution to remove the color.

(4) At this point, backflow is prone to occur. If this phenomenon happens, gently move the delivery tube to allow the liquid from the backflow to return to the receiving vessel. Then adjust the flame appropriately to restore normal reaction conditions.

【Thought questions】

(1) Why is a certain amount of water added when preparing bromoethane?

(2) What is the purpose of adding sulfuric acid to crude bromoethane?

(3) This experiment is prone to low yield, analyze the reasons.

5.1.2 Preparation of phenyl ethyl ether

【Experimental objectives】

To learn the synthesis of ethers by means of the Williamson method; To be familiar with the method of electromagnetic stirring.

【Experimental principle】

Ethers can be considered as compounds formed by the loss of one molecule of water between two alcohol molecules, or as derivatives of hydroxy compounds (such as alcohols, phenols, naphthols, etc.) in which a hydroxy hydrogen is replaced by an alkyl group. If the two groups in the ether are the same, the ether is called a simple or symmetrical ether; if the two groups are different, it is called a mixed or unsymmetrical ether. Simple ethers are usually prepared by the dehydration of alcohol molecules under acidic catalysis, while mixed ethers can be prepared by the reaction of alcohol salts and halogenated hydrocarbons.

$$C_6H_5OH + NaOH \longrightarrow C_6H_5ONa + H_2O$$

$$C_6H_5ONa + C_2H_5Br \longrightarrow C_6H_5OC_2H_5 + NaBr$$

【Chemicals】

Phenol	7.5 g (0.08 mol)
Sodium hydroxide	4 g (0.10 mol)
Bromoethane	12.4 g (8.5 mL, 0.12 mol)
Saturated sodium chloride aqueous solution	
anhydrous calcium chloride	

【Experimental procedure】

Install the device according to scheme 2-5. Place a magnet into a 250 mL three-neck RBF first, then 7.5 g phenol, 4 g NaOH, 4 g water are added separately. Add 8.5 g bromoethane into the dropping funnel.

Heat the mixture to dissolve the solid completely and start stirring. Heat gently to keep the temperature at 80～90°C. Bromoethane is added slowly by dropwise and it takes about 1 hr for the addition process. Aftermath, continue stirring for 1.5 hrs. Stop heating and stirring, add 10 mL water to dissolve the precipitate completely. Transfer the liquid into a separatory funnel, collect the organic layer. Wash the organic layer with saturated sodium chloride twice. Collect the crude product, dry with anhydrous calcium chloride.

Distillation. Collect the fraction of 165~173℃. When conducting vacuum distillation, you can refer to the pressure-boiling point relationship in the table below（Table 1）for distillation.

Table 1 The relationship of pressure-Boiling Point and Phenethyl Alcohol

Pressure/kPa	0.67	1.33	2.67	5.33	8.00	13.33	26.6
bp/℃	43.7	56.4	70.3	86.6	95.4	103.4	127.9

Data: colorless liquid, bp 170℃, d_4^{20} 0.966, n_D^{20} 1.5076.

【Thought questions】

What's the refluxing liquid during the reaction process? What's the solid generated? Why isn't the refluxing phenomenon obvious at the end of the reaction?

5.2 Series Experiment Ⅱ

5.2.1 Preparation of bromobenzene

【Experimental objectives】

To learn the halogenation reaction of aromatic compounds.

【Experimental principle】

The method for synthesizing aromatic halides is different from that of alkyl halides. Generally, halogens（chlorine or bromine）are used to react with aromatic compounds under the catalysis of iron powder or ferric halides. Through the electrophilic substitution reaction of aromatic hydrocarbons, the halogen atoms are directly introduced into the aromatic ring. The true catalyst is ferric halide. Initially, iron powder reacts with halogens to form ferric halides, which then act as catalysts. The entire substitution reaction proceeds as follows:

$$2Fe+3Br_2 \longrightarrow 2FeBr_3$$

$$FeBr_3+Br_2 \rightleftharpoons Br^+[FeBr_4]^-$$

$$C_6H_6 + Br^+ \rightleftharpoons [C_6H_6Br^+] \longrightarrow C_6H_5Br + H^+$$

$$[FeBr_4]^-+H^+ \longrightarrow FeBr_3+HBr$$

As the generation of halogen cations and tetrachloroferrate complex anions from ferric halides and halogens takes some time, there is an induction period before the halogen substitution reaction begins. For example, when synthesizing bromobenzene, the reaction is not noticeable at first but becomes intense after a certain period of time. During the reaction, it is necessary to slowly add bromine to excess benzene to avoid excessive reaction and reduce the production of byproduct dibromobenzene. Ferric halides are easily hydrolyzed, therefore, the reagents and instruments used must be anhydrous and dry.

Main reaction:

$$C_6H_6 + Br_2 \longrightarrow C_6H_5Br + HBr$$

Side reaction:

$$2\,C_6H_5Br + Br_2 \longrightarrow p\text{-}C_6H_4Br_2 + o\text{-}C_6H_4Br_2 + 2HBr$$

【Chemicals】

Benzene (anhydrous)	10 g (11.5 mL, 0.13 mol)
Bromine	16 g (5 mL, 0.1 mol)
Iron powder	0.25 g
Saturated sodium bisulfite	
Anhydrous calcium chloride	

【Experimental procedure】

In a dry 250 mL three-neck RBF, refluxing condenser, dropping funnel and stirring apparatus are installed. On the top of the condenser, a winding pipe is connected. The other side of the pipe is introduced into a conical flask with water. Make sure the mouth of the pipe is not immersed below the water surface. The conical flask is used to absorb hydrogen bromide.

In the RBF, 11.5 mL dry benzene, 0.25 g iron powder is added. 5 mL bromine is added into the dropping funnel.

At first, 1 mL bromine is added by dropwise into the RBF, the reaction sets in immediately (heat warmly if necessary), white fume is generated. The mixture is kept stirring while the remaining bromine is added dropwise. Keep the solution gently boiling for about 30 minutes. When all the bromine is added, continue heating the solution until no red bromine vapor or white fume evolves again. Let the mixture cool to room temperature, add 15 mL water to the RBF, keep stirring for about 2 minutes. Transfer the reactant to a separatory funnel to remove the aqueous layer. The crude bromobenzene is washed with saturated sodium bisulfite twice, then with water once. Separate the bromobenzene layer and dry it with anhydrous calcium chloride.

The dry crude bromobenzene is distilled to collect the part between 140~170℃. (note: an air condenser is used instead of a water condenser). The first part is distilled again to collect the part between 154~160℃.

Data: colorless liquid, bp 150℃, d_4^{20} 1.499, n_D^{20} 1.5597.

【Notes】

(1) This experiment can be combined with the "Preparation of Triphenylmethanol" experiment to form a series of experiments.

(2) The measurement of bromine should be carried out in a fume hood. Wear protective goggles and gloves, and be cautious not to inhale bromine vapor. In case of contact with bromine, wash with a solution of sodium bisulfite.

(3) Washing with water can remove iron bromide and also remove hydrogen bromide and some bromine.

(4) Washing with saturated sodium bisulfite solution can remove bromine. Since bromine has higher solubility in bromobenzene than in water, it is difficult to completely remove bromine using water.

【Thought questions】

(1) Which reagent is excess during the preparation of bromobenzene?

(2) Why is it necessary to use a dry flask, and what are the effects of water presence in benzene and bromine during the experiment?

5.2.2 Preparation of triphenylcarbinol

【Experimental objectives】

To learn Grignard Reaction.

【Experimental principle】

Alkyl halides react with metallic magnesium in anhydrous tetrahydrofuran or anhydrous diethyl ether to form alkyl magnesium halides, also known as Grignard reagents. The reaction of Grignard reagents with aldehydes, ketones, or epoxides is a common method for synthesizing various complex structured alcohols. The presence of water, oxygen, or other active hydrogen will destroy Grignard reagents, so the reaction must be carried out under anhydrous, oxygen-free, and non-active hydrogen conditions. The reaction can be carried out under an inert gas (such as helium or nitrogen) to effectively prevent Grignard reagents from reacting with oxygen and carbon dioxide. Coupling reactions are inevitable, but dilute solutions can be used to reduce the occurrence of such side reactions. Therefore, excessive alkyl halides should not be added before the reaction is truly initiated. Alkyl iodides are more prone to coupling reactions than alkyl bromides or alkyl chlorides, so alkyl bromides and alkyl chlorides with lower activity are preferred in the preparation of Grignard reagents.

The adducts formed by Grignard reagents and carbon-based compounds are hydrolyzed under acidic conditions. Since heat is released during hydrolysis, it should be carried out under cooling conditions. For alcohols that are easily dehydrated by acids, it is best to use saturated ammonium chloride solution for hydrolysis.

Grignard reagents can react with carboxylic acid esters to form addition products, which can be hydrolyzed to obtain tertiary alcohols.

$$C_6H_5MgBr + C_6H_5COOC_2H_5 \xrightarrow{ether} (C_6H_5)_2C(OC_2H_5)(OMgBr) \longrightarrow (C_6H_5)_2C{=}O + C_2H_5OMgBr$$

$$(C_6H_5)_2C{=}O + C_6H_5MgBr \xrightarrow{ether} (C_6H_5)_3C(OMgBr) \xrightarrow{NH_4Cl,\ H_2O} (C_6H_5)_3C\text{-}OH$$

【Chemicals】

Magnesium powder	1.5 g
Bromobenzene	10.49 g (7 mL, 0.068 mol)
Anhydrous ether	21.39 g (30 mL, 0.289 mol)
Ethyl benzoate	4.20 g (4 mL, 0.016 mol)
Ammonium chloride	7.5 g
Iodine	

【Experimental procedure】

Install the reaction device according to Scheme 2-6 and 2-7. a dry tube and an absorbing device are installed on the top of the condenser. Place 1.5 g magnesium, 1～2 particles of iodine inside a 250 mL three-neck RBF. In the dropping funnel, a mixture of 7 mL bromobenzene and 25 mL anhydrous ether is added. At first, about 5 mL solution is added by dropwise inside the RBF. The reaction sets in quickly. The color of iodine disappears gradually. If it doesn't react, heat the RBF gently to induce the reaction. When the reaction sets in, start stirring and add the remaining solution into the RBF. Keep the reaction gently boiling all the time. After the addition is over, keep boiling for another half an hour until the magnesium disappears completely.

Cool the RBF with an ice-water bath. A mixture of 4 mL ethyl benzoate and 5 mL anhydrous ether is added into the dropping funnel. It is added into the rbf while stirring by dropwise. Keep refluxing to 0.5 hr after the addition is over.

Cool the RBF with an ice-water bath. In the dropping funnel, 30～35 mL of saturated ammonium chloride is added. It works to decompose the addition product. Transfer the reactant to a separatory funnel, ether layer is collected and transferred to a 100 mL RBF. It is distilled to remove ether which is recycled later. Finally, steam-distillation is carried out to get rid of unreacted bromobenzene and side product biphenyl. The distillation is continued until no oily material is out. Inside the RBF, the product appears in a solid state. The crude product is recrystallized from ethanol-water. White crystal is obtained.

Data: mp 164.2°C.

【Notes】

(1) This experiment must be conducted under anhydrous conditions, and all reaction instruments and reagents must be thoroughly dried before use.

(2) The magnesium shavings used should be freshly prepared. If there is an oxide layer on the surface, it can be removed by using 5% hydrochloric acid for several minutes, and then washed successively with water, ethanol, and ether, and dried under vacuum.

(3) Due to the difficulty of Grignard reaction with bromobenzene, a small amount of iodine crystals are added to initiate the reaction.

(4) To prevent excessive generation of byproducts due to overly vigorous reaction, the addition of the mixture of bromobenzene and diethyl ether should not be too fast.

(5) To prevent the saturated solution of ammonium chloride from decomposing and losing its effectiveness, the laboratory uses the latest preparation, which involves dissolving 7.5 g of ammonium chloride in 32 mL of water.

(6) Since the process is an exothermic reaction, the addition of the reagent NH_4Cl should be slow, otherwise, the diethyl ether may be ejected due to excessive reaction. If there are still flocculent magnesium hydroxide residues that have not completely dissolved during the reaction, a small amount of dilute hydrochloric acid solution can be added until they dissolve completely.

【Thought questions】

(1) Can bromobenzene be added too quickly or all at once when preparing the Grignard reagent? What are the consequences?

(2) What are the side reactions that occurred during the preparation of the Grignard reagent in this experiment? How can they be avoided?

(3) Instead of using the decomposition of saturated ammonium chloride as the additive in this experiment, what other reagents can be used?

(4) Why is it necessary for the instruments and reagents to be dried in each step before hydrolysis in this experiment? What methods should be employed?

(5) What measures are taken to initiate the reaction when preparing phenylmagnesium bromide, and are there any other methods?

(6) When is it appropriate to add activated charcoal during recrystallization using a mixed solvent? Which solvent should be used for washing the crystals?

5.3 Series Experiment Ⅲ

5.3.1 Preparation of nitrobenzene

【Experimental objectives】

To learn the nitration of benzene, be familiar with the usage of air condenser.

【Experimental principle】

Aromatic nitro compounds are generally prepared by direct nitration of aromatic compounds. Different nitrating agents are used according to the activity of the substrate. A mixture of concentrated nitric acid and concentrated sulfuric acid, also known as mixed acid, is one of the commonly used nitrating agents. Nitration is an irreversible reaction. The role of concentrated sulfuric acid in mixed acid is not only for dehydration but also to promote the formation of nitronium cations (NO_2^+), thereby increasing the reaction rate.

Nitrobenzene is the simplest and most readily available aromatic nitro compound. It is synthesized in both laboratory and industrial production by reacting benzene with mixed acid at around 50℃. If the reaction temperature is too high, dinitrobenzene may be formed easily.

$$\text{Main reaction: } C_6H_6 + HONO_2 \xrightarrow[40\sim50℃]{H_2SO_4} C_6H_5NO_2 + H_2O$$

$$\text{Side reaction: } C_6H_5NO_2 + HONO_2 \xrightarrow{H_2SO_4} m\text{-}C_6H_4(NO_2)_2 + H_2O$$

【Chemicals】

Benzene	23.7 g（27 mL, 0.28 mol）
Nitric acid	30 mL（*d*=1.40, 0.67 mol）
Concentrated sulfuric acid	38 mL（*d*=1.84, 0.71 mol）
Sodium carbonate soln.	10%
Saturated sodium chloride soln.	
Anhydrous calcium chloride	

【Experimental procedure】

38 mL concentrated sulfuric acid is added slowly into a conical flask containing 30 mL concentrated nitric acid. The conical flask is put inside an ice-water bath. The mixed acid is placed at a standstill to cool down to room temperature.

Install the device according to Scheme 2-6. In a 250 mL three neck RBF, 27 mL benzene is added. Start stirring. Transfer the mixed acid into a dropping funnel and start the addition process. Cool the RBF with cold water in order to keep the reacting temperature within 40～50°C. After the addition is over, heat the mixture to 60°C and keep at 0.5 h.

When the mixture is cold, separate the acid layer with a separatory funnel and recycle the mixed acid. The crude nitrobenzene is washed with water twice and then with 10 mL 10% Na_2CO_3 until the organic layer is neutral. Pour the crude nitrobenzene into a dry conical flask and dry it with anhydrous calcium chloride.

Pour the clear nitrobenzene into a 50 mL RBF. 1～2 particles of zeolite are added. Connect an air condenser and start to distil. Collect the part of 205～210°C. Please be cautious not to distil completely.

Data: colorless（pain yellow）liquid, smell of almond, bp 210.9°C, d_4^{20}=1.203, n_D^{20}= 1.5562.

【Notes】

（1）The experiment can be part of a series of experiments with "Preparation of Aniline" and "Preparation of Acetanilide".

（2）Nitration reaction is an exothermic reaction. When the temperature exceeds 50℃, a significant amount of dinitrobenzene is generated, and some nitric acid and benzene are evaporated.

（3）The nitration of benzene is an exothermic reaction. When adding the mixed acid, it is important to add it in small portions because the nitration reaction is fast. As the mixed acid is added and dinitrobenzene is formed, the amount of benzene in the reaction mixture gradually decreases, slowing down the nitration rate. Therefore, after adding half of the mixed acid, the amount added each time can be increased.

(4) Take a small amount of the upper layer of the reaction mixture, drop it into saturated NaCl solution, and when the oil droplets sink, it indicates that the nitration reaction is basically complete.

(5) Industrial concentrated sulfuric acid often contains impurities such as mercury salts, which act as catalysts and result in the production of trace amounts of poly-nitrophenols, such as picric acid and dinitrophenol, giving the alkaline solution a yellow color.

(6) If the impurities of nitric acid in dinitrobenzene are not washed away, they will decompose during the final distillation process, generating NO_2 gas and increasing the possibility of dinitrobenzene formation.

(7) Nitrobenzene is toxic, so it should be handled with great care. If it splashes on the skin, it can be wiped with a small amount of alcohol and then rinsed with soap and warm water.

(8) The crude dinitrobenzene after washing is cloudy due to the presence of small water droplets. After adding a drying agent, it can be heated with a water bath (30~50℃) and shaken to accelerate drying, then cooled and left for distillation.

(9) The final distillation temperature should not exceed 210℃, and it should never be distilled to dryness to avoid explosions.

【Thought Questions】

(1) What would be the result if the freshly prepared mixed acid was added to benzene all at once?

(2) What is the reason for controlling the reaction temperature between 40~50℃ in the experiment?

(3) What is the purpose of sequentially washing the crude product with water, alkaline solution, and water? How do they differ in reaction conditions?

(4) Why should the residue in the boiling flask not be distilled to dryness during the distillation purification process?

5.3.2 Preparation of aniline

【Experimental objectives】

To learn the reduction reaction of aromatic compounds and steam distillation.

【Experimental principle】

The preparation of aromatic amines cannot be achieved by directly introducing an amino group (—NH_2) onto the aromatic ring. Instead, it is usually prepared by an indirect method. Reducing aromatic nitro compounds in acidic medium is an important method for preparing aromatic amines.

$$ArNO_2+6[H] \longrightarrow ArNH_2+2H_2O$$

Common reducing agents include iron-hydrochloric acid, iron-acetate, tin-hydrochloric acid, stannous chloride-hydrochloric acid, etc. When preparing aniline from nitrobenzene, iron filings are commonly used in a solution containing a small amount of acidic electrolyte.

$$C_6H_5NO_2 \xrightarrow{Fe/HCl} C_6H_5NH_2$$

【Chemicals】

Nitrobenzene 6.02 g（5 mL, 0.05 mol）

Iron powder 15 g（0.27 mol）

Concentrated hydrochloric acid 1 mL

NaCl

Na_2CO_3

NaOH

【Experimental procedure】

Install the device according to Scheme2-6（please note the mercury ball should immerse below the surface）. In the three neck RBF, 25 mL water, 15 g iron powder, 1 mL concentrated hydrochloric acid and 5 mL nitrobenzene are added. 1～2 particles of zeolite are added at the same time. Start stirring and heat the mixture to a boil. Because the reaction is exothermic, remove the heating device as soon as the mixture begins refluxing. When the reaction becomes mild, continue heating for another 30 minutes until the mixture turns black, which shows the reaction is complete.

When the mixture is cooled down to room temperature, add 1 g Na_2CO_3 powder while shaking until the mixture turns basic.

The device is modified into a steam distillation device（according to Scheme 2-11）and starts the steam distillation. When the distillate is not turbid, stop. Add NaCl into the distillate until saturated. Separate the aqueous layer with a separatory funnel. Pour the organic layer to a dry conical flask and dry it with solid NaOH.

Pour the dry crude aniline to a 50 mL RBF and start to distill, collect the part of 182～185℃.

Data: colorless oily liquid, bp. 184.4℃, d_4^{20}=1.022, n_D^{20}=1.5863.

【Inspection and Testing】

（1）Reaction of Aniline with Bromine Water: Add a drop of aniline to 5 mL of water, stir until it dissolves, then take 5 mL of the aniline water solution and add saturated bromine water. Immediately, a white turbid substance or precipitate will form.

（2）Amines have alkaline properties and can form water-soluble salts with strong acids. This can be used to identify amines. Add a drop of aniline to 5 mL of water and stir until it dissolves. Add 6 drops of concentrated hydrochloric acid, stir, and observe the phenomenon. Finally, dilute the solution with water, and it should remain transparent.

【Notes】

（1）This experiment uses concentrated hydrochloric acid as a catalyst, but it can also be replaced with 1 mL of glacial acetic acid or 0.25～1 g of ammonium chloride.

（2）Both aniline and nitrobenzene are toxic, so caution should be exercised to avoid skin contact or inhalation of their vapors. If accidentally in contact with the skin, rinse with water first and then wash with soap and warm water.

(3) This step is mainly to activate the iron and shorten the reduction time.

(4) When the reaction has been ongoing for about 20 minutes, the reactants turn brown, and the color gradually darkens, eventually becoming black, indicating that the reaction is complete. To check if the reaction is complete, a small amount of the reaction mixture can be sucked up with a pipette and dropped into 1 $mol \cdot L^{-1}$ hydrochloric acid. If no oil droplets are observed, it indicates that the reaction is complete.

(5) The reaction mixture should be cooled before adding sodium carbonate powder to prevent foam overflow.

(6) After the completion of the reaction, if there is a sticky black-brown substance on the walls of the three-necked flask, it can be removed by heating with a 1∶1 (volume ratio) solution of hydrochloric acid and water.

(7) At 22°C, approximately 3 mL of aniline dissolves in every 100 mL of water. To reduce the loss of aniline, according to the principle of salting out, salt is added to saturate the distillate. Add finely ground NaCl at a rate of 20~25 g per 100 mL of distillate. As a result, the majority of the aniline originally dissolved in water will precipitate as an oily substance.

(8) Due to the high boiling point of aniline, air condensers can be used for distillation.

【Thought questions】

(1) Why is it necessary to thoroughly stir the reaction mixture during the reaction process?

(2) Why is sodium carbonate used to make the reaction mixture alkaline before performing steam distillation? Which test paper is used to check?

(3) Why can crude aniline be separated using steam distillation?

(4) How can nitrobenzene be removed if it is present in the synthesized aniline?

(5) Why is solid sodium hydroxide used for drying before distillation and purification?

5.3.3 Preparation of acetanilide

【Experimental objectives】

To learn the acetylation of aniline.

【Experimental principle】

In organic synthesis, acetylation of aromatic amines is often used to "protect" the amino group, reducing the sensitivity of aromatic amines to oxidizing reagents and appropriately reducing the reactivity of the aromatic ring to facilitate the generation of mono-substituted products. Due to the steric effect of the acetyl group, substitution occurs mainly at the para position. After the reaction is completed, the acetyl group is removed by hydrolysis.

The reaction between aniline and acetic acid is a reversible reaction. In order to prevent the hydrolysis of acetylaniline and improve the yield, this experiment uses the method of excess reactant (acetic acid) and continuously removing the product (water) from the reaction to shift the equilibrium to the right. Since the boiling point of water is close to that of acetic acid (bp 117.9℃), a fractionating column is used, and the top temperature of the fractionating column is strictly controlled below 110℃ to effectively remove the by-product water.

$$C_6H_5NH_2 + CH_3COOH \rightleftharpoons C_6H_5NHCOCH_3 + H_2O$$

【Chemicals】

Aniline 5.1 g (5 mL, 0.055 mol)

Glacial acetic acid 8.3 g (8 mL, 0.132 mol)

Zinc powder

activated charcoal

【Experimental procedure】

Install the reacting device according to Scheme 2-10.

5 mL newly distilled aniline and 8 mL glacial acetic acid are added into the RBF. Zinc powder is added later. Heat to boil gently and keep the temperature between 105℃ and 110℃ to distil the water generated in the reaction completely. About 40～60 minutes later, stop heating.

Transfer the reactant into a beaker containing 100 mL water while stirring. Keep stirring and cool the beaker in order to precipitate the crude acetanilide completely. Collect the solid with suction, wash the solid with cold water to remove the excess acid.

Transfer the crude acetanilide to a beaker, add 80 mL water, heat to boil to dissolve all the solid. 0.5 g charcoal is added while stirring. Continue heating to boil for 5 minutes. Hot filtration is carried out immediately. The filtrate is cooled down to room temperature. Acetanilide precipitates as plate crystal. Collect with suction. Remove the excess water as much as possible. Dry the crystal on a watch glass.

Data: white plate crystal, mp 114℃.

【Testing and Inspection】

(1) Under acidic conditions, the amide derivatives undergo hydrolysis and are then treated with sodium nitrite to form diazonium salts. These salts can react with β-naphthol to produce a precipitate ranging from bright red to dark brown in color.

$$RCONHAr \xrightarrow{H^+} RCOOH + ArNH_3^+ \xrightarrow{NaNO_2} ArN_2^+ \xrightarrow{\beta\text{-萘酚}} \text{1-(Ar-N=N)-2-naphthol}$$

Take a test tube and add about 0.2 g of acetanilide, followed by 1 mL of 70% sulfuric acid. Heat the mixture until the acetanilide is completely dissolved. Then cool the solution to 5℃ and add 3～4 drops of 5% sodium nitrite aqueous solution, followed by a few drops of 10% NaOH solution of 5% β-naphthol. Observe for any color changes. The formation of a precipitate ranging from bright red to dark brown indicates the presence of acetanilide.

(2) Infrared spectroscopic identification: Obtain the infrared spectrum of the product and compare it with the reference spectrum shown below for identification purposes.

Vibration frequencies in the infrared spectrum of acetanilide include the following peaks:

3200～3300 cm^{-1}	N—H Strecthing vibration
1670 cm^{-1}	C=O stretching vibration
1560 cm^{-1}	N—H bending vibration
760 cm^{-1}	C—H of benzene bending vibration
699 cm^{-1}	C—H of benzene bending vibration

【Notes】

(1) Benzene amine, if stored for a long time, tends to oxidize and become dark in color. It should be distilled before the experiment.

(2) Zinc powder is mainly used to prevent the oxidation of benzene amine during the reaction. Care should be taken not to add too much, otherwise it will form zinc hydroxide which is not conducive to subsequent treatment.

(3) Oil droplets refer to molten water-containing acetanilide. The solubility of acetanilide in 100 mL of water is: 25℃, 0.563 g; 80℃, 3.5 g; 100℃, 5.2 g.

(4) To prevent supersaturation of the solution, adding a certain amount of water is beneficial for hot filtration.

【Thought questions】

(1) What should be paid attention to in this experiment to ensure complete reaction?

(2) What should be taken into consideration during the recrystallization process to improve yield and quality?

第六部分

有机化学设计性实验

6.1 设计性实验总体要求

设计性实验要求学生运用已学习过的知识，通过查阅文献，借鉴他人的经验，设计出常量或半微量或微型实验的方案，在教师认可后并在教师指导下，自己动手合成某些有实用价值的中间体或化合物。也可以结合教师科学研究的需要，合成一些原料或中间体，通过设计实验进一步培养学生的综合能力，培养学生独立进行研究和创新的能力。

学会查阅文献，要注意的是，各文献的实验步骤和条件往往不同，有些内容出于保密等原因而不翔实，这就要求学生能运用已获得的知识和技能独立地进行正确的判断、综合。通过透彻掌握目标分子的合成原理、主副反应、产物（含副产物）的有关性能（如溶解度、熔点、沸点），设计出可行的实验方案（包括合成路线、使用的原料与试剂、仪器的选用、操作条件的控制、主副产物的分离、产品的精制、鉴定等）。

设计方案经教师审定后，学生独立进行实验。实验用量最好半微量，也可进行微型实验。

实验后除了要交出产品，还应写出设计实验报告。设计实验报告格式可参照期刊论文，应当包括题目、作者、提要（摘要）、关键词、实验内容、结果讨论、主要参考文献等栏目，要简要地介绍题目的背景、实验的目的意义，要有实验步骤的精确描述（包括原料的配比和用量、工艺流程和实验条件、有关数据和现象等），要有实验结果的有关数据（包括产物的产量和收率、产品有关物理参数及文献值、图表、波谱及其他有关数据），要有讨论（包括对实验结果的评价、对实验的改进意见、意外情况的分析及自己的心得体会等）。

教师在设计性实验实施的过程中要始终起指导作用。对各设计性实验必须都心中有数，为学生独立完成实验提供必要的软硬件支持。在审查学生制定的设计方案时，切不可忽略对安全因素的审查。在实验过程中要观察和评价学生的操作技术正确与否，必要时要及时予以纠正。要随时准备解答学生实验中出现的各种问题。总之，在充分体现学生的主体作用的同时，也要充分发挥教师的主导作用。

6.2 双酚A的合成

（一）提示

1. 双酚 A 的化学名称是 2,2-双（4'-羟基苯基）丙烷。该化合物是一个用途很广的化工原料。它是双酚 A 型环氧树脂及聚碳酸酯等化工产品的合成原料，还可用作聚氯乙烯塑料的热稳定剂，电线防老剂，油漆、油墨等的抗氧剂和增塑剂。

2. 从文献看，双酚 A 的制备方法主要是通过苯酚和丙酮的缩合反应：

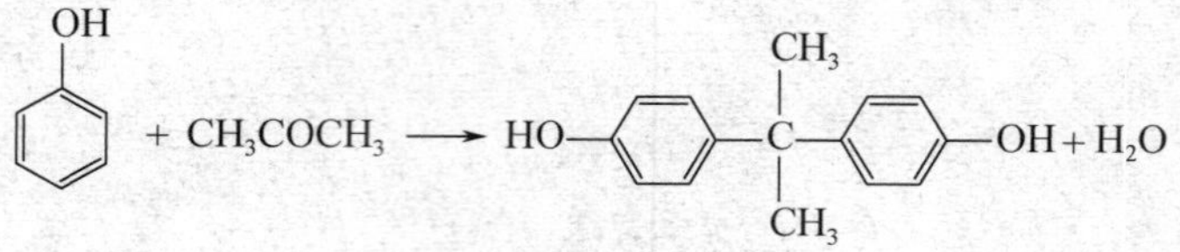

$$C_6H_5OH + CH_3COCH_3 \longrightarrow HO-C_6H_4-C(CH_3)_2-C_6H_4-OH + H_2O$$

反应在 CCl_4、$CHCl_3$、CH_2Cl_2、C_6H_5Cl 等有机溶剂中进行。盐酸、硫酸等质子酸作催化剂。

3. 还有其他的合成方法。

（二）要求

1. 查阅有关文献，设计并确定一种可行的制备实验方案。
2. 制备 2～5 g 的双酚 A 产品。

6.3 苄叉丙酮的合成

（一）提示

1. 苄叉丙酮的化学名称是 4-苯基-3-丁烯-2-酮。它是一种用途广泛的有机化合物，尤其是在香料工业和电镀工业中。它本身是肉桂醛香料系列中的一个香料，以它为原料得到的一系列衍生物也很重要；在电镀工业中，它和其他成分一起被配成溶液，用作一些合金的光亮剂，例如铅锡合金、铅锌合金的光亮剂。除此以外，它还具有一定的杀虫活性和驱虫功效。能用作杀虫剂中的稳定剂。

2. 主要合成方法

a. 醛酮缩合反应

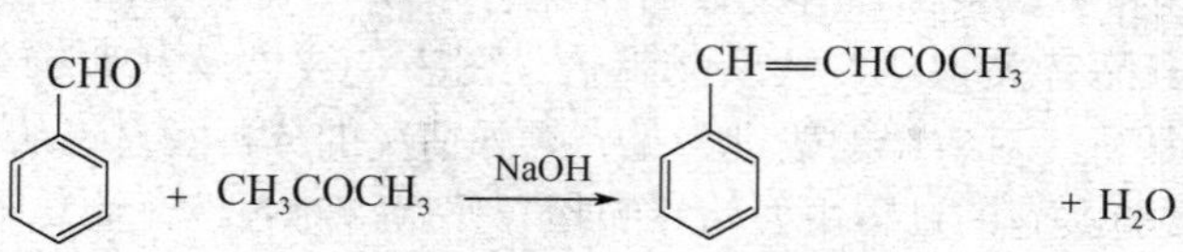

$$C_6H_5CHO + CH_3COCH_3 \xrightarrow{NaOH} C_6H_5CH{=}CHCOCH_3 + H_2O$$

b. 与酸化试剂反应

$$C_6H_5CHO + (CH_3CO)_2O \xrightarrow[LiClO_4]{CH_3COOH} C_6H_5CH{=}CHCOCH_3 + CH_3COOH$$

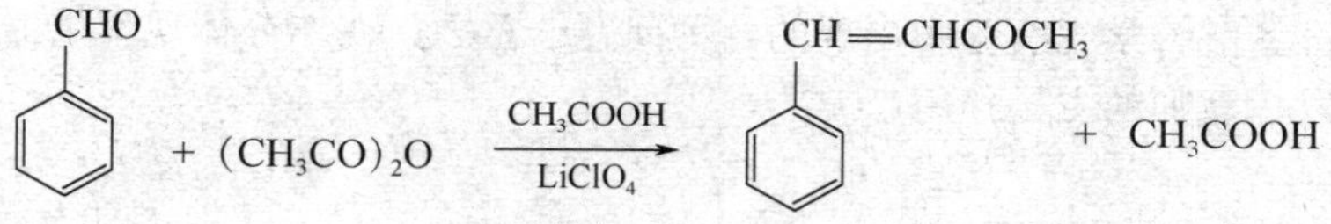

此外，还有其他合成方法。

（二）要求

1. 查阅相关文献，设计并确定一种切实可行的实验方案（最好是半微量或微型实验）。
2. 合成 0.5～2 g 左右的产品。
条件许可前提下，可以探讨各合成方法的特点。

6.4 乙酰基二茂铁的合成

（一）提示

1.二茂铁是一个非常稳定的化合物。它的化学稳定性是由于分子中最高成键轨道和最低反键轨道之间的能垒较大（约为 4 eV），同时也由于轨道之间成键时交盖较好。Fe^{2+}中心原子的所有最外层价键轨道（3d，4s，4p）与两个环戊二烯的 p 轨道交叠，使得二茂铁比游离的环戊二烯负离子具有更大的“芳香性”，因此二茂铁可以发生典型的亲电取代反应，进行 Friedel-Crafts 反应的相对难易程度与其他化合物比较如下：

$$PhOH \approx PhFePh > PhOCH_3 > PhOsPh > PhH$$

所以，从二茂铁制备乙酰基二茂铁容易进行。

2. 该化合物制备的基本反应

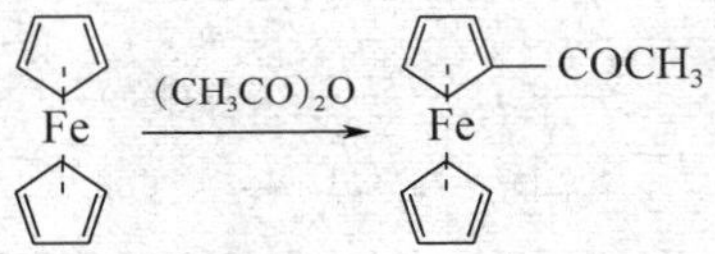

$(CH_3CO)_2O$ Fe → Fe — $COCH_3$

3. 反应粗产物可以用柱色谱分离法进行提纯。

（二）要求

1.查阅相关文献，设计并确定一种可行的半微量制备或微型实验方案。
2.合成产品 0.5～2 g，测定产品熔点。
3.注意催化剂的选用。

6.5 Diels-Alder 环加成反应

（一）提示

1. 共轭二烯与带有强吸电子基烯烃的环加成反应很容易进行，甚至需要冰水冷却以控制反应平稳地进行。环戊二烯与顺丁烯二酸酐的反应可表达为：

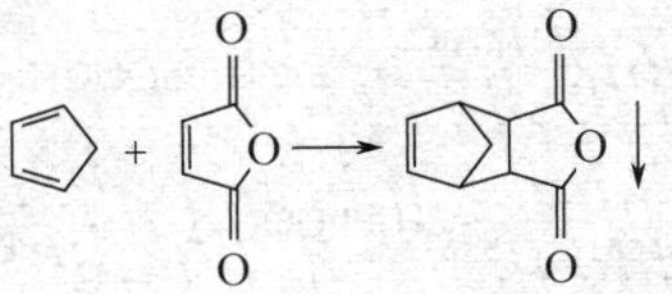

O O O → O O O ↓

2. 亦可以用呋喃与顺丁烯二酸酐的环加成反应进行设计实验。

（二）要求

1.查阅文献后，设计出合理可行的半微量制备或微型实验方案；
2.制备 0.5～2 g 产品，测定其熔点。

6.6 席夫碱及其铜配合物合成

（一）提示

1. 人们为了研究生物体内蛋白质与过渡金属离子配合物所产生的生命活动，常合成一些结构类似但又简单的配合物，从对这些模拟化合物的研究中可以观察到类似的生命现象。合成 Co-salen 或 Cu-salen 就是这项研究工作的一部分。Salen 是水杨醛（邻羟基苯甲醛）与乙二胺缩合形成的产物——席夫碱，通常将醛类与有机胺类缩合的产物统称为席夫碱。

2. 合成的基本反应是：

$$(o\text{-HO})C_6H_4CHO + H_2NCH_2CH_2NH_2 \xrightarrow{EtOH} (o\text{-HO})C_6H_4CH{=}NCH_2CH_2N{=}CHC_6H_4(o\text{-OH})$$

$$\xrightarrow{Cu^{2+}/CH_3COO^-} \text{Cu-salen}\ (\text{N-CH}_2\text{-CH}_2\text{-N},\ \text{CH},\ \text{OH} \rightarrow \text{Cu} \leftarrow \text{OH})$$

（二）要求

1. 设计出合理的半微量或微型实验方案。
2. 制备 0.5～2 g 的 salen。
3. 制备 0.5～2 g 的 Cu-salen。

6.7 以甲苯为原料的三步合成

（一）提示

1. 这个设计实验是多步合成的训练。例如，以甲苯为原料，通过 Friedel-Crafts 酰基化反应制得 p-$CH_3PhCOCH_3$ 再与苯甲醛进行 Claisen-Schmidt 缩合得到 α，β-不饱和酮 p-$CH_3PhCOCH{=}CHPh$（Z，E 两种），进而与溴进行 1,2-和 1,4-亲电加成，从而制得最终产物：

$$PhCH_3 + (CH_3CO)_2O \xrightarrow{AlCl_3} p\text{-}CH_3PhCOCH_3 + CH_3COOH$$

$$p\text{-}CH_3PhCOCH_3 + PhCHO \xrightarrow[CH_3CH_2OH]{NaOH} \underset{H}{\overset{p\text{-}CH_3PhCO}{}}C{=}C\underset{Ph}{\overset{H}{}} + \underset{H}{\overset{p\text{-}CH_3PhCO}{}}C{=}C\underset{H}{\overset{Ph}{}}$$

2. 鉴于多步合成的产品越来越少，制定方案时可以先做常量实验，而后是半微量实验，最后是微型实验。

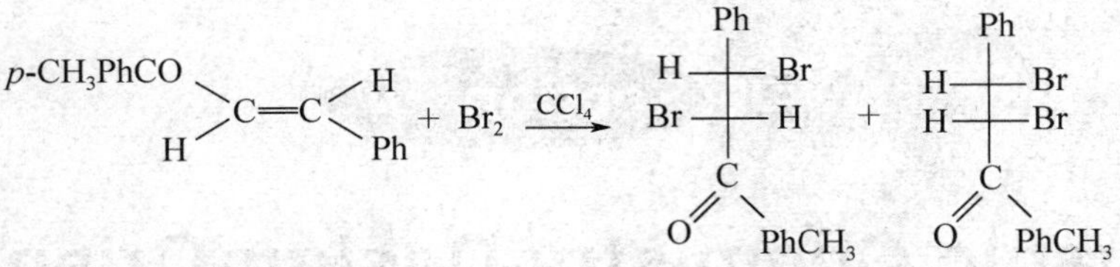

$$p\text{-}CH_3PhCO(H)C{=}C(H)Ph + Br_2 \xrightarrow{CCl_4} Ph\text{-}CHBr\text{-}CHBr\text{-}C(=O)PhCH_3 + Ph\text{-}CHBr\text{-}CHBr\text{-}C(=O)PhCH_3$$

（二）要求

1. 在制定完整的多步合成设计实验方案之后，原则上应按方案进行实验。

2. 如果由于主客观原因，自制的产品（或得不到产品）不可能继续进行下一步实验，允许用实验室提供的原料（如果可能的话）继续进行实验。但实验室不得提供多于两步合成的原料（含初始原料）。

3. 各步合成的产品均应由老师验收。

4. 最终产品产量不少于 0.1 g，并应有质量检测数据。

6.8 苯甲酸乙酯的制备

要求

（1）合成路线按苯甲醇→苯甲酸→苯甲酸乙酯的顺序进行；

（2）写出各步反应式；

（3）列出仪器药品，画出每步装置图；

（4）写出实验步骤；

（5）制定检验鉴定方法。

基准量：0.02 mol 原料。

Section VI

Organic Chemistry Design-Oriented Experiments

6.1 Overall requirements for experimental design

These Design-oriented experiments require students to apply the knowledge they have learned, consult literature, draw on the experiences and lessons of predecessors, and design plans for constant or semi-micro or micro experiments. After approval by the teacher and under the guidance of the teacher, students are expected to synthesize certain intermediates or compounds of practical value by their own hands. Students can also synthesize some raw materials or intermediates according to the needs of teachers' scientific research, further cultivating students' comprehensive abilities and their ability to independently conduct research and innovation.

Students should learn to consult literature, However, it is important to note that the experimental steps and conditions recorded in various literatures are often different from each other, and some contents are not detailed due to confidentiality reasons. This requires students to independently make correct judgments and synthesis based on the knowledge and skills they have acquired. By thoroughly understanding the synthesis principles of target molecules, main and side reactions, and relevant properties of products (such as solubility, melting point, boiling point), students should design feasible experimental plans (including synthesis routes, selection of raw materials and reagents, selection of instruments, control of operating conditions, separation of main and side products, purification and identification of products, etc.).

After the design plan is approved by the teacher, students independently carry out the experiments. It is preferable to use small quantities or conduct micro-scale experiments.

In addition to submitting the products, students should also write a report on the design experiment. Its format can refer to the papers in general chemistry and chemical engineering journals, and should include sections such as title, authors, abstract, keywords, experimental content, result discussion, and main references. It should briefly introduce the background and purpose of the topic, provide accurate descriptions of experimental steps (including raw material proportions and quantities, process flow and experimental conditions, relevant data and phenomena, etc.), present relevant data on experimental results (including product yield and recovery, physical parameters of product quality compared to literature values, charts, spectra, and other related data, etc.), and include discussions (including evaluation of experimental results, suggestions for improving the experiment, analysis of unexpected situations, personal insights and reflections).

Teachers should always play a guiding role in the process of designing and implementing experiments. They should have a clear understanding of each design experiment and provide necessary software and hardware support for students to independently complete the experiments. When reviewing the design plans made by students, it is crucial not to overlook the examination of safety factors and ensure absolute safety. During the experiment process, teachers should observe and evaluate whether students' operational techniques are correct or not, and make timely corrections if necessary. They should be prepared to answer various questions that may arise during students' experiments. In conclusion, teachers should fully embody the students' active role while fully exerting their own leading role.

6.2 Synthesis of Bisphenol A

1. Tips

(1) Bisphenol A, also known as 2, 2-bis (4-hydroxyphenyl) propane, is a widely used chemical raw material. It is a starting material for the synthesis of bisphenol A epoxy resin and polycarbonate, and can also be used as a heat stabilizer for PVC plastic, antioxidant and plasticizer for paints, inks, etc.

(2) According to the literature, bisphenol A is mainly prepared by the condensation reaction of phenol and acetone: The reaction is carried out in organic solvents such as CCl_4, $CHCl_3$, CH_2Cl_2, and C_6H_5Cl. Proton acids such as hydrochloric acid and sulfuric acid are used as catalysts.

(3) There are other synthetic methods available.

2. Requirements

(1) Consult relevant literature, design and determine a feasible preparation experiment plan.

(2) Preparation of 2-5 g of bisphenol A product.

6.3 Synthesis of Benzylideneacetone

1. Tips

(1) Benzylideneacetone, also known as 4-phenyl-3-buten-2-one, is a widely used organic

compound, especially in the fragrance and electroplating industries. It is itself a fragrance in the cinnamaldehyde fragrance series, and a series of derivatives obtained from it are also important. In the electroplating industry, it is mixed with other components to form a solution used as a brightener for some alloys, such as brighteners for lead-tin and lead-zinc alloys. In addition, it also has certain insecticidal activity and deworming effect, which can be used as a stabilizer for insecticides.

(2) Main synthetic methods: a. Aldol condensation reaction

$$C_6H_5CHO + CH_3COCH_3 \xrightarrow{NaOH} C_6H_5CH{=}CHCOCH_3 + H_2O$$

b. Reaction with oxidizing agents

$$C_6H_5CHO + (CH_3CO)_2O \xrightarrow[LiClO_4]{CH_3COOH} C_6H_5CH{=}CHCOCH_3 + CH_3COOH$$

In addition, there are other synthetic methods available.

2. Requirements

(1) Consult the relevant literature, design and determine a feasible experimental plan (preferably semi-micro or micro).

(2) Synthesize about 0.5~2 g of product. Under permitted conditions, the characteristics of each synthesis method can be discussed.

6.4 Synthesis of Acetylferrocene

1. Tips

(1) Ferrocene is a very stable compound. Its chemical stability is due to the large energy barrier (about 4 eV) between the highest bonding orbital and the lowest antibonding orbital in the molecule, as well as the good overlap of orbitals during bonding. All the outermost valence bond orbitals of the Fe^{2+} center atom (3d, 4s, 4p) overlap with the p orbitals of two cyclopentadienyl rings, making ferrocene more "aromatic" than free cyclopentadienyl anion. Therefore, it is relatively easy to carry out typical electrophilic substitution reactions and Friedel-Crafts reactions on ferrocene, compared with other compounds such as PhOH≈PhFePh > $PhOCH_3$ > PhOsPh > PhH.

(2) The basic reaction for preparing this compound.

$$Fe(C_5H_5)_2 \xrightarrow{(CH_3CO)_2O} (C_5H_5)Fe(C_5H_4COCH_3)$$

(3) The crude product of the reaction can be purified by column chromatography.

2. Requirements

(1) Consult relevant literature, design and determine a feasible semi-micro or micro experimental plan.

(2) Synthesize 0.5~2 g of the product and measure its melting point.

(3) Pay attention to the selection of catalysts.

6.5 Diels–Alder Cycloaddition Reaction

1. Tips

(1) The cycloaddition reaction between conjugated dienes and electron-deficient dienophiles is easily carried out and may even require cooling with ice water to control the reaction smoothly. The reaction between cyclopentadiene and maleic anhydride can be expressed as follows:

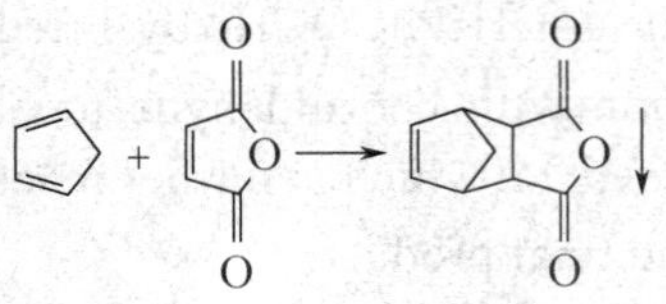

(2) Alternatively, the cycloaddition reaction between furan and maleic anhydride can be used for experimental design.

2. Requirements

(1) After reviewing the literature, design a reasonable and feasible semi-microscale or microscale experimental procedure.

(2) Prepare 0.5-2 grams of product and determine its melting point.

6.6 Synthesis of Schiff Bases and Their Copper Complexes

1. Tips

(1) In order to study the biological activities of protein-metal ion complexes in living organisms, it is common to synthesize structurally similar but simpler complexes. By studying these simulated compounds, similar biological phenomena can be observed. The synthesis of Co-salen or Cu-salen is part of this research. Salen is the product formed by condensing salicylaldehyde (2-hydroxybenzaldehyde) with ethylenediamine, and the products formed by condensing aldehydes with organic amines are commonly referred to as salen.

(2) The basic reaction for synthesis is as follows:

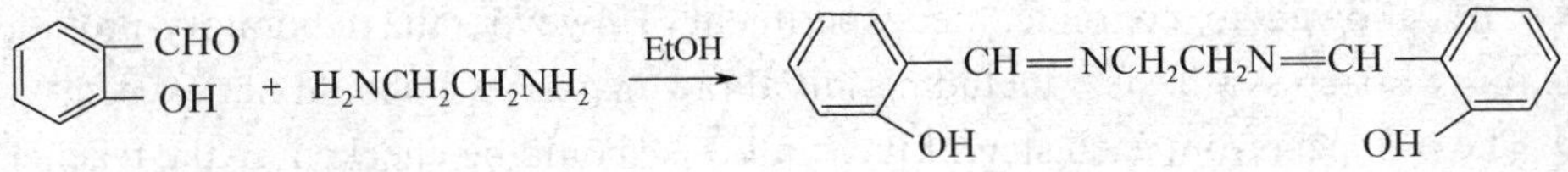

$$\text{C}_6\text{H}_4(\text{CHO})(\text{OH}) + H_2NCH_2CH_2NH_2 \xrightarrow{\text{EtOH}} \text{C}_6\text{H}_4(\text{OH})\text{—CH}{=}NCH_2CH_2N{=}\text{CH—C}_6\text{H}_4(\text{OH})$$

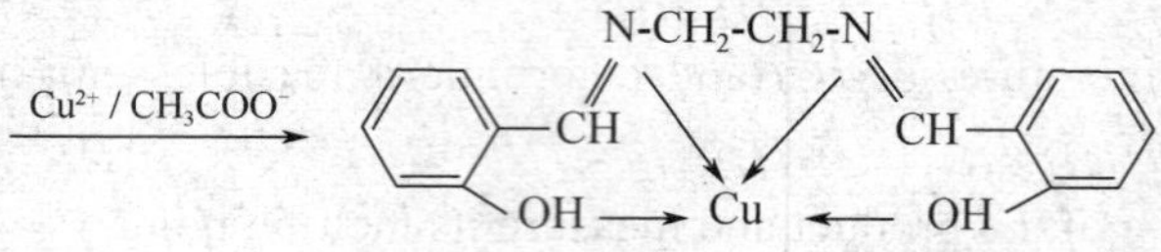

$\xrightarrow{Cu^{2+} / CH_3COO^-}$ N-CH_2-CH_2-N / CH CH / OH → Cu ← OH

2. Requirements

(1) Design a reasonable semi-microscale or microscale experimental procedure.

(2) Prepare 0.5～2 grams of salen.

(3) Prepare 0.5～2 grams of Cu-salen.

6.7 Three-step Synthesis using Toluene as Raw Material

1. Tips

(1) This design experiment is a training for multi-step synthesis. For example, toluene is used as the raw material to obtain *p*-$CH_3PhCOCH_3$ by Friedel-Crafts acylation reaction, and then Claisen-Schmidt condensation with benzaldehyde to obtain α, β-unsaturated ketone *p*-$CH_3PhCOCH{=}CHPh$ (*Z*, *E* isomers), which is further reacted with bromine in 1, 2-and 1, 4-electrophilic addition to obtain the final product:

$$PhCH_3 + (CH_3CO)_2O \xrightarrow{AlCl_3} p\text{-}CH_3PhCOCH_3 + CH_3COOH$$

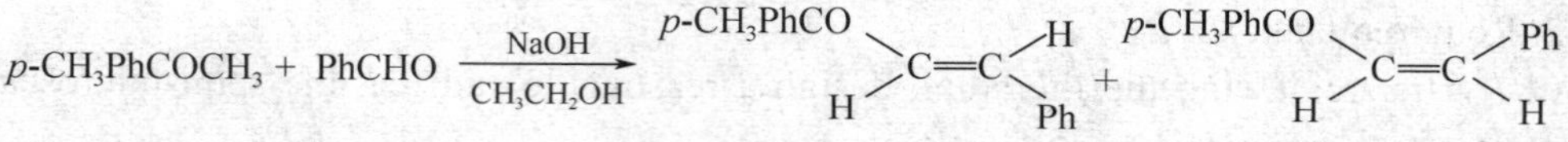

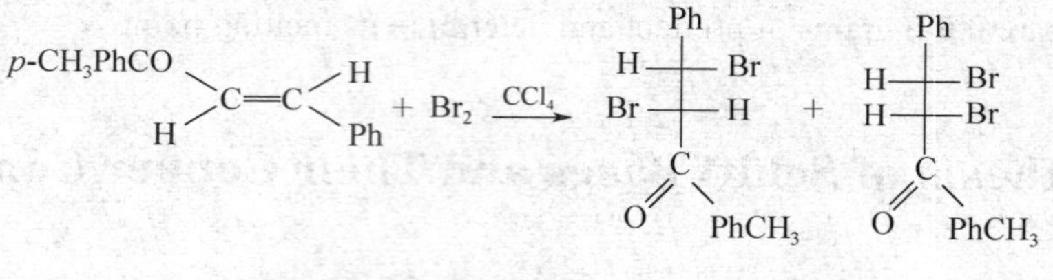

$$p\text{-}CH_3PhCOCH_3 + PhCHO \xrightarrow[CH_3CH_2OH]{NaOH} (p\text{-}CH_3PhCO)(H)C{=}C(H)(Ph) + (p\text{-}CH_3PhCO)(H)C{=}C(Ph)(H)$$

$$(p\text{-}CH_3PhCO)(H)C{=}C(H)(Ph) + Br_2 \xrightarrow{CCl_4} Ph\text{-}CH(Br)\text{-}CH(Br)\text{-}C(=O)PhCH_3 + Ph\text{-}CH(Br)\text{-}CH(Br)\text{-}C(=O)PhCH_3$$

(2) Given that the yield of multi-step synthesis products tends to decrease, constant experiments should be conducted first, followed by semi-microscale experiments, and finally microscale experiments when developing the complete multi-step synthesis experimental plan.

2. Requirements

(1) After developing a complete multi-step synthesis experimental plan, experiments should be conducted according to the plan.

(2) If, due to subjective or objective reasons, self-made products cannot continue to the next step of the experiment (or cannot be obtained), laboratory-provided raw materials (if possible) are allowed to continue the experiment. However, the laboratory-provided raw materials for two-step synthesis (including initial raw materials) should not exceed.

(3) The products from each step of the synthesis should be checked by the teacher.

(4) The final product yield should be no less than 0.1 g and there should be quality

testing data.

6.8 Preparation of Ethyl Benzoate

Requirements

(1) The synthesis route should follow the sequence of benzyl alcohol → benzoic acid → ethyl benzoate.

(2) Write down the reaction equations for each step.

(3) List the instruments and reagents, and draw the apparatus diagram for each step.

(4) Write down the experimental procedures.

(5) Develop testing and identification methods.

Reference amount: 0.02 mol of starting material.

参考文献

【1】邢其毅，裴伟伟，徐瑞秋，等.基础有机化学.北京：北京大学出版社，2017.

【2】陈荣业.有机反应机理解析与应用.北京：化学工业出版社，2017.

【3】宁永成.有机化合物结构鉴定与有机波谱学.4 版.北京：科学出版社，2018.

【4】徐继有.有机合成安全学.北京：科学出版社，2016.

【5】赵摇兴，孙祥玉.有机分子结构光谱鉴定.2 版.北京：科学出版社，2018.

【6】Jie Jack Li，有机人名反应——机理及合成应用：第 5 版.荣国斌译.北京：科学出版社，2021.

【7】邵国成，张春艳.实验室安全技术.北京：化学工业出版社，2016.

【8】曹静、陈星、孙圣峰.化学实验室安全教程.北京：化学工业出版社，2023.

【9】张奇涵.有机化学实验.3 版.北京：北京大学出版社，2015.

【10】冒爱荣，吴玉芹.有机化学实验.北京：化学工业出版社，2021.

【11】孟长功.基础化学实验.3 版.北京：高等教育出版社，2019.

【12】王箴.化工辞典.4 版.北京：化学工业出版社，2005.

【13】李景宁.有机化学.6 版.北京：高等教育出版社，2018.

【14】王积涛.有机化学.3 版.天津：南开大学出版社，2018.

【15】胡宏纹.有机化学.5 版.北京：高等教育出版社，2020.

【16】李明.有机化学实验.3 版.北京：高等教育出版社，2019.

【17】王俊儒.有机化学实验.3 版.北京：高等教育出版社，2019.

【18】赵建庄.有机化学实验.2 版.北京：中国林业出版社，2018.

【19】姚刚，王红梅.有机化学实验.2 版.北京：化学工业出版社，2018.